Molecular Recognition: Chemical and Biochemical Problems

Special Publication No. 78

# Molecular Recognition: Chemical and Biochemical Problems

The Proceedings of an International Symposium

University of Exeter, April 1989

Edited by
**Stanley M. Roberts**
Department of Chemistry
University of Exeter

ROYAL
SOCIETY OF
CHEMISTRY

**British Library Cataloguing in Publication Data**
Molecular recognition.
1. Molecules. Interactions. Biochemical aspects
I. Roberts, Stanley M. (Stanley Michael)
574.19'2

ISBN 0-85186-796-0

Published by the Royal Society of Chemistry, Thomas Graham House, Science Park,
Cambridge CB4 4WF

Printed in Great Britain by Henry Ling Ltd., at the Dorset Press, Dorchester, Dorset

# Preface

Molecular recognition is a much-used expression which means different things to different scientists. The interaction of two simple molecules involves mutual recognition; the interaction of two macromolecules (e.g. DNA/protein) would fall into the same category. The articles in this book mainly fall centre stage, describing work involved in determining small molecule/large molecule interactions.

The interactions of an enzyme with a substrate is a typical form of such molecular recognition. A detailed understanding of such recognition is often the starting point for the discovery of novel pharmaceutical agents based on the inhibition of a key enzyme in a pathogenic organism. Enzyme inhibition is also an important feature of the new compounds coming forward for use in agricultural chemistry.

The design of new molecules for specific tight-binding to a selected protein is at the forefront of the list research objectives within the area of molecular recognition. Not only are novel enzyme inhibitors required but also substances are sought that interact in a selective and selected way with certain receptors, virally coded protein, and DNA.

Synthetic chemistry is vitally important in this area of work for a number of reasons. First the compounds that are designed to be enzyme inhibitors are often complex with carefully positioned functional groups within the molecules: these compounds have to be made by multistep organic synthesis. The synthesis of peptides and (deoxy) ribonucleotides is also of great

v

importance in this field of research.

The use of powerful spectroscopic methods is a key feature in much of the work. The employment of n.m.r. has been most popular and this technique has provided great insight into the regions of the small molecule/large molecule complex that are in close proximity. Molecular graphics has allowed modelling of the pairs of interacting molecules and can give clues as to how the 'recognition' can be improved.

Several other matters recur in the following articles. A number of authors highlight the importance of understanding the energetics of intra- and inter-molecular interactions using molecular mechanics calculations which are necessarily becoming increasingly sophisticated. It is also obvious that in the process of molecular recognition and the consequent molecular interactions the solvent, often water, must be considered to play a part. Displacement of the solvent from the space between the interacting partners must be taken into account.

Another area of concern has been the ordering of molecules at a metal surface. Processes in this category are of importance in the understanding of the mechanism of action and improvement of heterogeneous catalysts. The design of new, useful conductors is possible.

Studies in the area of molecular recognition often benefit from the collaboration of biochemists, chemists (organic, physical, inorganic, and theoretical) and molecular biologists. It is 4/5 years since the area of molecular recognition itself received recognition for special funding in Europe, the USA and Japan. Progress has been good in many areas and spectacular in some cases. Yet it is still a developing field and many of the following Chapters reflect the excitement of the new thinking and indicate possible points of breakthrough in the next few years.

The authors are a noteworthy group of world experts in the area of molecular recognition. Their efforts have provided a stimulus for others and the articles are thought-provoking summaries of challenging work. Very few of the Chapters describe complete packages of research; most illustrate the starting points for their major studies in the area and the ideas that are

described may provide an inspiration for other
scientists.

Stanley M Roberts

# Contents

# Molecular Recognition and Drug Design

By Simon F. Campbell

PFIZER CENTRAL RESEARCH, SANDWICH, KENT, UK

## 1.    Introduction

Molecular recognition processes control every aspect of life on our planet. Thus, the ability of individual molecules to recognise, and discriminate between, closely related partners is a key determinant of chemical reactivity, enzyme catalysis, gene regulation and many other fundamental processes. Understanding molecular recognition processes, particularly those which influence the binding of small molecules to complex proteins, is an essential skill for medicinal chemists, and is a vital component of modern drug discovery. New drugs must be highly discriminating at the molecular level so that diseases can be controlled without untoward side effects. Therefore, drug molecules must be specifically designed to recognise only those particular enzymes or receptors involved in the disease process, and to ignore the multitude of close relatives which subserve normal biological mechanisms. Moreover, the medicinal chemist should also be aware that in vivo degradation can offset intrinsic potency and receptor/enzyme selectivity, and interaction of drug molecules with metabolising enzymes must be minimised.

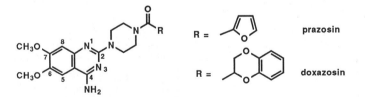

This account summarises our understanding of structure activity relationships for a series of 2,4-diamino-6,7-dimethoxyquinazoline[3] which display high affinity and selectivity for $\alpha_1$-adrenoceptors.[1-6] Agents of this mechanistic class, such as prazosin[7] and doxazosin[8], are now widely available for the treatment of hypertension. In addition, these drugs produce potentially beneficial effects on plasma lipids, which may be important in reducing the risks of heart disease.[9]

## 2.    Prazosin, the prototype $\alpha_1$-adrenoceptor antagonist

Prazosin was synthesised in 1965 in our Groton laboratories during a search for novel antihypertensive agents with vasodilator properties. Animal and clinical evaluation showed that prazosin was an effective, safe antihypertensive agent which reduced blood pressure without increasing heart rate. At the time however, it was difficult to rationalise this unique pharmacological profile in terms of a specific mechanism of action. Although prazosin displayed affinity for $\alpha$-adrenoceptors, the compound was quite distinct from classical $\alpha$-antagonists, and alternative mechanisms were indirectly implicated. However, in the 1970s it was demonstrated that, in addition to the $\alpha_1$-adrenoceptors on the blood vessel wall, a second subtype ($\alpha_2$) was present on the sympathetic nerve endings.[10] It was soon shown that prazosin was a potent, selective antagonist of the $\alpha_1$-mediated vasoconstrictor actions of noradrenaline but did not interfere with the prejunctional $\alpha_2$-sites which modulate transmitter release.[11,12] By contrast, phentolamine was non-selective, whereas yohimbine showed some preference for $\alpha_2$-adrenoceptors. These studies provided a compelling rationale for the clinical profile observed with prazosin and for the poor antihypertensive efficacy of earlier $\alpha$-antagonists. The demonstration of absolute discrimination between $\alpha_1$- and $\alpha_2$-adrenoceptors by prazosin was an important stimulus in re-awakening interest in the field, not only in our own laboratories. Thus, we initiated a new research programme with the objective of identifying second generation $\alpha_1$-antagonists with potential advantages over prazosin. However, in order to place our synthetic programme on a rational basis we decided first to define the structural features and molecular recognition processes which underwrote the exceptional potency and selectivity demonstrated by prazosin for $\alpha_1$-adrenoceptors.

## 3.    Recognition of the quinazoline nucleus by the $\alpha_1$-adrenoceptor

Our approach to defining molecular recognition processes derived from the fact that prazosin (1) is a potent ($pA_2$=8.37 $\pm$ 0.24) competitive antagonist of the $\alpha_1$-mediated responses to noradrenaline (2), and from the subsequent assumption that these molecules compete for common receptor binding sites. Indeed, considerable structural similarity does exist between prazosin and noradrenaline and it might be expected that $\alpha_1$-antagonist activity could be expressed in less complex analogues. Even so, the high affinity and selectivity displayed by 4-amino-2-dimethylamino-6,7-dimethoxyquinazoline (3) for $\alpha_1$-adrenoceptors is quite remarkable, and clearly demonstrates that the quinazoline nucleus present in prazosin dominates receptor interactions (Fig. 1).

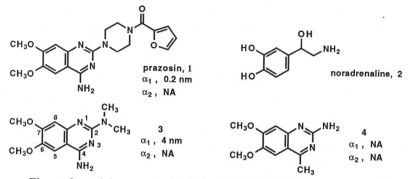

Figure 1. α-Adrenoceptor binding affinity ($K_i$ values) for representative quinazoline derivatives.[13] (NA indicates no activity at $10^{-6}$M)

Moreover, the enthalpy of binding for 3 (-11.49 kcal/mole) at $\alpha_1$-sites is greater than predicted by the Andrews' approach[14] (-9.3 kcal/mole) which confirms a particularly effective receptor fit, and suggests quite specific molecular recognition processes. Elimination of one or both of the 2-amino substituents in 3 reduced $\alpha_1$-affinity by some 10- and 50-fold respectively whilst removal of the 6,7-dimethoxy groups totally abolished activity. Introduction of any substituents into the quinazoline nucleus which reduced basicity such that protonation was unfavoured at physiological pH (7.4), also obliterated $\alpha_1$-affinity (e.g. 4, pKa = 5.2). This latter observation was not unexpected since noradrenaline is also highly basic (pKa = 9.6) as are the vast majority of other α-agonist/antagonist structures. These initial SAR studies suggested strongly that the protonated 2,4-diamino-6,7-dimethoxyquinazoline nucleus present in prazosin (pKa = 6.8 ± 0.04) and 3 (pKa = 8.1 ± 0.08) might serve as a particularly effective, conformationally-restricted bioisostere for noradrenaline.

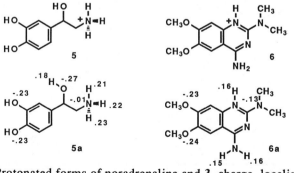

Figure 2. Protonated forms of noradrenaline and 3, charge localised (5,6); delocalised (5a,6a); CNDO/2 charge distribution by Mulliken population analysis (only selected centres shown).

Noradrenaline contains only a single basic centre and the protonated species (5) will predominate (<u>ca</u> 95%) at physiological pH. At first glance, four potential protonation sites are available for 3 (ca. 80% protonated at physiological pH) although molecular orbital calculations show that N-1 is overwhelmingly favoured over the exocyclic nitrogen centres (6) Similar conclusions have been reached for quinazoline dihydrofolate reductase inhibitors[15] whilst protonation and quaternisation of 2,4-diaminopyrimidine derivatives also predominate at N-1.[16]

These results suggest major electronic differences between the amino function of noradrenaline and the quinazoline 2-nitrogen atom and their ability to participate in the same molecular recognition processes must be questioned. However, formal location of positive charge on single nitrogen centres (5,6) is solely a matter of convenience since molecular orbital calculations indicate considerable dispersion over neighbouring atoms.[17,18] Thus, for noradrenaline most of the positive charge resides on the three hydrogen atoms attached to the nitrogen (5a), and there seems little reason to focus on either the location, or Coulombic interaction, of an essentially neutral nitrogen centre in agonist-receptor recognition. Charge delocalisation is more extensive for the protonated quinazoline (6a) and, although both the N-1 H and 4-NH$_2$ functions could act as focal points in any receptor recognition process, only the former will be considered initially. These results suggested that charge-reinforced hydrogen bonding would be important for both agonist and antagonist molecules, providing an anionic site is present on the receptor which is equally accessible to both noradrenaline and the quinazoline series. Computer-simulated superimposition of 5a and 6a (Fig. 3) shows how these two molecules might compete for the same receptor site which could contain a hydrophobic area to accommodate the aromatic rings from either series, a recognition site for the vicinal oxygen atoms and an anionic centre (A) which accepts a positively charged hydrogen atom from either protonated species.

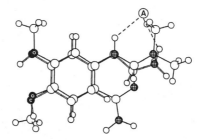

**Figure 3.** Computer simulated superimposition of 5a (hollow bonds, benzylic hydroxy function removed for clarity) and 6a (solid bonds) with respect to common anionic centre A.

In order to refine this simple model, we next made two further assumptions:- (a) that the aromatic ring, the quinazoline N-1 H and the counterion were coplanar and (b) that the anion should be capable of binding simultaneously with the benzylic hydroxyl and the ammonium head of noradrenaline. In order to identify potential receptor counterions, the interaction of chloride, phosphate and carboxylate anions with noradrenaline was evaluated using molecular mechanics techniques to identify favourable binding positions followed by full relaxation energy minimisation to optimise interaction geometries. Final binding energies (enthalpies of interaction) were calculated by standard INDO methods. Although all three counterions appeared equally well suited for detailed evaluation,[19,20,21] the carboxylate anion was selected mainly because salt bridges between aspartate and glutamate residues and protonated heterocyclic nucleii have been detected in other enzyme systems.[22] Interestingly, the amino acid sequences of several receptor subtypes have been determined in the last few years, and a common aspartate residue has been proposed as an important recognition centre for the 'onium heads of various natural transmitters.[23,24]

For noradrenaline, a coplanar cyclic hydrogen bonding arrangement (Fig. 4, 7) is preferred (binding energy, -155.93 kcal/mol), a conformation which lies close (<2 kcal/mol) to the global minimum (phenyl ring rotated through 60°) and which would easily be accessible to the natural $\alpha_1$-transmitter. Similarly, a protonated quinazoline derivative also demonstrated charge-reinforced hydrogen bonding (8, binding energy, -72.2 kcal/mol) but closer approach (<2.5Å) of the carboxylate anion was prevented by the piperidine ring.[25]

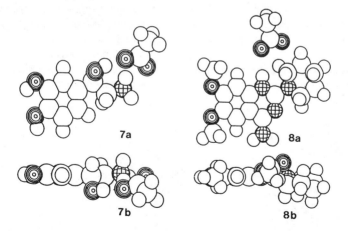

**7a**

**8a**

**7b**

**8b**

**Figure 4.** Interaction of noradrenaline (**7**) and a quinazoline derivative (**8**) with a carboxylate counterion. Face-on (a) and side-views (b) illustrated.

Comparison of structures (7) and (8) indicates that the carboxylate counterions are differently located (ca. 4Å separation) relative to the parent aromatic rings and appear incompatible with the previous concept of fixed recognition sites. However, while initial agonist and antagonist receptor recognition may be similar, the consequences of binding are quite different. Agonists activate the receptor to produce a physiological response whereas antagonists exert "squatters' rights" and need not disturb the active site. Indeed, for both $\alpha$- and $\beta$-receptors, agonist binding is enthalpy-driven, consistent with strong bonding to the receptor in order to overcome an unfavourable decrease in entropy.[26] On the other hand, antagonist interaction is entropy-driven, since an important contribution to binding affinity results from the entropy increase associated with released water molecules.[27]

Complex 8 may therefore represent the interaction of the protonated heterocycle with the ground state of the $\alpha_1$-adrenoceptor, and the high binding affinity reflects a hydrophobic attraction, charge-reinforced hydrogen bonding and a favourable entropy component associated with release of bound water molecules.[28] This drug-receptor complex is also a minimum enthalpy arrangement, given the original constraints, and any conformational reorganisation required for receptor activation would be energetically unfavourable.

When noradrenaline approaches this receptor ground state, the carboxylate counterion forms a hydrogen bond with the benzylic hydroxyl function and initiates a medium-range electrostatic interaction with the ammonium head (Fig. 5, 9 binding energy, -74.54 kcal/mol). Charge-reinforced hydrogen bonding can then be optimised by a 4Å migration of the counterion (7, binding energy, -155.93 kcal/mol). Thus, the conformational change in the protein structure usually associated with receptor activation could be promoted by the free energy decrease accompanying transformation of the initial agonist complex (9) into the more stable arrangement (7). A simple $\alpha_1$-adrenoceptor model can therefore be proposed which rationalises the different consequences of agonist and antagonist receptor occupancy.

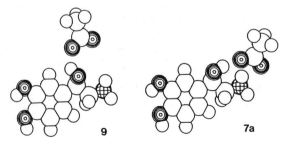

9                                             7a

**Figure 5.** Interaction of noradrenaline with a carboxylate counterion; $\alpha_1$-adrenoceptor ground state (9); activated state (7a).

Importantly, for S(+)- noradrenaline, which is 100-fold less potent than the natural transmitter, the counterion complex is less favoured (binding energy, -141.86 kcal/mol) due to repulsive interactions between the hydroxyl proton and the positively charged ammonium head. Moreover, the remarkable $\alpha_1$-adrenoceptor selectivity of this quinazoline series is consistent with a receptor model containing the hydrophobic binding area and carboxylate counterion in parallel planes. By contrast, initial analysis of $\alpha_2$-adrenoceptor SARs suggests that these key areas may be orthogonal to one another,[29] and a flat quinazoline nucleus could not accommodate such alternative geometry.

A key feature of the above receptor model was the suggestion that an N-1 protonated quinazoline was exquisitely suited for charge reinforced hydrogen bonding with a carboxylate counterion in the ground state conformation of the $\alpha_1$-adrenoceptor. In order to substantiate these proposals, we decided to replace the parent nucleus by isosteric heteroaromatic systems such as quinoline (**10**) and isoquinoline (**11**). N-1 protonation to provide the required pharmacophore is only possible for **10** and comparison of these isomeric series provides a critical test for previous modelling studies. The 2,4-diaminoquinoline ring system (**10**) was constructed *via* a novel intramolecular cyclisation of an acetamidine derivative under basic or Lewis acid conditions.[30]

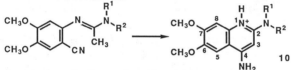

Entry into the isoquinoline series (**11**), albeit in moderate yield, was effected by treatment of 2-methyl-4,5-dimethoxybenzonitrile with LDA at -70°C followed by addition of an appropriate cyanamide.[31]

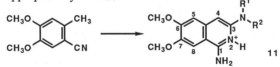

In general, a wide range of 2,4-diaminoquinoline derivatives showed similar or higher, $\alpha_1$-adrenoceptor affinity to the original quinazoline series whereas the corresponding isoquinolines were weak or inactive.[30,31] These results are best illustrated by comparison of the data in Table I. Thus, in the quinoline series, the 2-dimethylamino analogue (**12**) is roughly half as potent as **3** whereas **13** displayed similar activity to prazosin. Thus, these quinoline and quinazoline systems appear to be recognised in a common fashion at the $\alpha_1$-adrenoceptor with N-1 presumably playing a similar role for either nuclei. Moreover, the enhanced basicity of these 2,4-diaminoquinolines compared to the corresponding quinazolines allows more facile protonation at N-1. Indeed, at physiological pH, **13** will exist mainly (86%)

**Table 1.** Binding[13] and pKa Data for Quinazoline, Quinoline and Isoquinoline Derivatives

| No. | X | Y | $R_1,R_2$ | $\alpha$-receptor binding affinity[a], $K_i$, (nM) | pKa |
|-----|---|---|-----------|--------------------------------|-----|
| 3 | N | N | $(CH_3)_2$ | $4.10 \pm 0.62$ | $8.1 \pm 0.08$ |
| prazosin | N | N | $(CH_2CH_2)_2NCO$-2-furyl | $0.19 \pm 0.02$ | $6.8 \pm 0.04$ |
| 12 | N | CH | $(CH_3)_2$ | $11.37 \pm 2.00$ | $9.3 \pm 0.09$ |
| 13 | N | CH | $(CH_2CH_2)_2NCO$-2-furyl | $0.14 \pm 0.07$ | $8.18 \pm 0.03$ |
| 14 | CH | N | $(CH_3)_2$ | NA | $7.1 \pm 0.09$ |
| 15 | CH | N | $(CH_2CH_2)_2NCO$-2-furyl | $160 \pm 29$ | - |

[a] apart from prazosin ($K_i$, $4830 \pm 1280$ nM for displacement of [$^3$H]clonidine), none of these compounds displayed $\alpha_2$-adrenoceptor affinity up to $10^{-6}$M

as the N-1 protonated form whereas only 20% protonation of prazosin will occur. Functional assays show that **13** is a highly potent ($pA_2 = 9.76 \pm 0.26$) competitive antagonist of the $\alpha_1$-mediated vasoconstrictor effects of noradrenaline and is some 20 times more potent than prazosin. Thus, the enhanced basicity of **13** may be more evident in functional, rather than binding, assays since the former requires efficient displacement of the noradrenaline cation.

By contrast, the isoquinoline (**14**) shows no relevant affinity for $\alpha_1$-adrenoceptors even though substantial protonation (34%) would be expected at physiological pH. However, protonation of **14** will occur on N-2, as confirmed by X-ray analysis of the hydrochloride salt of **15**.[31] Comparison of the positive charge distribution in the protonated forms of **3,12,14** shows that electron densities on the dimethoxy, amino and dimethylamino functions are very similar, although, obviously, the ring protonation sites are quite different (Table 2). These results provide strong support for the proposal that N-1 protonation is a fundamental requirement for effective interaction of these heterocyclic nucleii with the $\alpha_1$-adrenoceptors and that other functionalities may be of secondary importance. For example, an alternative receptor binding mode in which the primary amino function acts as a bioisostere for the benzylic hydroxyl group in noradrenaline appears most unlikely.

**Table 2.** Calculated Positive Charge Distribution in the Protonated Forms of **3,12,14** (CNDO/2 Mulliken population analysis, only selected centres shown).

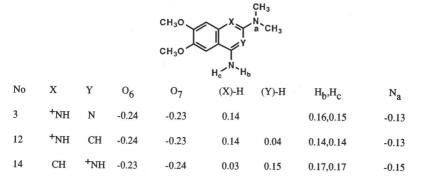

| No | X | Y | $O_6$ | $O_7$ | (X)-H | (Y)-H | $H_b,H_c$ | $N_a$ |
|---|---|---|---|---|---|---|---|---|
| 3 | $^+$NH | N | -0.24 | -0.23 | 0.14 | | 0.16,0.15 | -0.13 |
| 12 | $^+$NH | CH | -0.24 | -0.23 | 0.14 | 0.04 | 0.14,0.14 | -0.13 |
| 14 | CH | $^+$NH | -0.23 | -0.24 | 0.03 | 0.15 | 0.17,0.17 | -0.15 |

The modest $\alpha_1$-adrenoceptor affinity exhibited by **15** probably results from a hydrophobic interaction involving the 3-substituent since extrapolation of the pKa data in Table 1 shows that the molecule would not be efficiently protonated at physiological pH. The differences in binding affinity between prazosin and **15** (binding energies, -13.3 and -9.3 kcal/mol) indicate that charge-reinforced hydrogen bonding between the N-1 protonated quinazoline nucleus and an anionic site on the receptor contributes about 4.0 kcal/mol. However, computer-assisted comparison of the X-ray structures of prazosin and **15** shows that the piperazino moieties are displaced from one another, although, obviously the parent heterocyclic nuclei are an exact match. Rotation of the piperazine ring into a coplanar arrangement with the isoquinoline nucleus allows an almost exact fit with prazosin, albeit at a cost of some 1.0 - 1.6 kcal/mol. If this coplanar arrangement of **15** is a prerequisite for recognition at the $\alpha_1$-adrenoceptor then the binding energy between the N-1 protonated quinazoline nucleus and the carboxylate counterion on the protein can be revised to 2.4 - 3.0 kcal/mol. This value is quite close to a recent estimate (1.8 kcal/mol) for the binding free energy of salt bridge formation between a protonated pteridine nucleus and an aspartate anion in dihydrofolate reductase[32].

In conclusion, these studies provide strong support for the $\alpha_1$-adrenoceptor model proposed previously and confirm the importance of the N-1 protonated quinazoline and quinoline pharmacophores for effective interaction with the receptor active site. The molecular recognition processes which contribute to the exceptionally high binding affinity of these systems appear to be quite specific and are exquisitely dependent on charged reinforced hydrogen bonding. It is quite remarkable that relatively simple structures such as **3** and **12** can compete most effectively with noradrenaline at the receptor active site, and that a subtle difference in protonation site can have a devastating effect on the biological activity of **14**.

## 4.    The role of the quinazoline 2-substituent

The SAR analysis so far has stressed the important role of the quinazoline nucleus in prazosin in dominating receptor interactions, and it is not immediately obvious whether an extended 2-substituent provides any additional benefits. Inspection of the data in Table 3 indicates a thousand fold increase in binding affinity for prazosin over the unsubstituted 2-amino analogue (**16**) although a more detailed analysis is required to establish "goodness of fit". Thus, while compounds **16** and **17** may be the weakest members of this quinazoline series, the observed enthalpies of binding are some 1.5 - 1.7 kcal/mole higher than expected from Andrews calculations. These simple compounds are therefore exceptionally well-tailored for the $\alpha_1$-adrenoceptor although the receptor fit can be further optimised with **3**.

**Table 3** Binding affinities[13], binding energies[14], and antihypertensive activities[34] for representative quinazoline derivatives

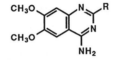

| No. | R | $\alpha_1$-receptor binding affinity ($K_i$nM) | binding energy (kcal/mole) obs | calc | % reduction in SHR blood pressure |
|---|---|---|---|---|---|
| 16 | NH₂ | 190 | 9.21 | 7.7 | 5 |
| 17 | NHCH₃ | 37±15.0 | 10.18 | 8.5 | - |
| 3 | N(CH₃)₂ | 4.1±0.62 | 11.49 | 9.3 | - |
| 18 | N-piperidine | 6.1±1.2 | 11.25 | 11.7 | 18 |
| 19 | N-piperidine-phenyl | 1.0 | 12.33 | 15.2 | 26 |
| 20 | N-piperidine-O⌒O⌒ | 1.81±1.45 | 11.97 | 13.6 | 83 |
| 21 | N-piperidine-CONHBu | 0.67±0.2 | 12.57 | 16.7 | 100 |
| 22 | N-piperazine-N-phenyl | 3.4 | 11.60 | 15.6 | 45 |
| prazosin | N-piperazine-N-C(O)-furan | 0.19±0.02 | 13.32 | 18.6 | 70 |

Elaboration of the 2-dimethylamino moiety in **3** has little immediate effect on binding affinity (**18-20**) although 10- and 20-fold improvements are achieved with **21** and prazosin. However, there are now substantial discrepancies between the observed and calculated binding energies and "goodness of fit" of these more elaborate molecules has obviously deteriorated. Moreover, the wide variation of physicochemical properties (CLOG P3 values[33]) of the quinazoline 2-substituents (data not shown) seems to have little influence on molecular recognition processes at the $\alpha_1$-adrenoceptor. These studies suggest that the quinazoline 2-substituent (R) may occupy a relatively open site on the receptor, and that any improvements in potency derive from the entropy gain as water molecules are forced from the active site, rather than from any specific contact with the protein structure. These observations, coupled with the high binding affinity displayed by **16,17** and **3**, reinforce the concept that the N-1 protonated quinazoline/quinoline nuclei are particularly effective bioisosteres for noradrenaline which participate in highly efficient molecular recognition processes at the $\alpha_1$-adrenoceptor. Finally, the selectivity of the compounds in Table 3 for $\alpha_1$- rather than $\alpha_2$-adrenoceptors was at least 1,000 and, in most cases, was substantially greater (data not presented[1-6]).

Although binding studies provide a most convenient measure of intrinsic receptor affinity, potential drugs must be able to block the functional effects of noradrenaline. Indeed, all of the compounds in Table 3, which were tested, proved to be potent, competitive antagonists of the $\alpha_1$-mediated, vasoconstrictor actions of noradrenaline. Moreover, these compounds, like prazosin, did not interfere with the prejunctional $\alpha_2$-adrenoceptor which modulates transmitter release.[1-6]

However, while this quinazoline series obviously demonstrates outstanding potency and selectivity for $\alpha_1$-adrenoceptors in vitro, SARs for antihypertensive activity in vivo must also be defined since high receptor affinity affords no protection against metabolic vulnerability, poor oral absorption or limited pharmacokinetics. The quinazoline derivatives in Table 3 were therefore evaluated in the spontaneous hypertensive rat (SHR) since this model is sensitive to most clinically effective antihypertensive agents, and also allows a fairly rapid compound throughput. It is immediately apparent from the data in Table 3 that, while the quinazoline 2-substituent has some influence on binding affinity, it plays a major role in governing in vivo performance. Thus, compound **16** is weakly active, and only a modest improvement is observed with the cyclised derivatives **18,19**. However, introduction of an appropriate substituent into the piperidine ring (**20,21**) has a marked impact on antihypertensive efficacy which is maintained with an N-acylpiperazino derivative such as prazosin itself. In addition to absolute reductions in blood pressure, duration of action is also important since once-daily administration of antihypertensive agents is preferred in clinical practice. Thus, more extensive evaluation[4-6] showed that the antihypertensive activities of **21** and prazosin were maintained over the whole test period (4.5h) in SHR whereas the response to 20 was obviously waning. These results demonstrate that the quinazoline 2-substituent plays a key role in influencing antihypertensive activity and duration of action, and that appropriate structural modification would be an important feature in the design of superior analogues.

### 5.    Novel, clinically effective $\alpha_1$-adrenoceptor antagonists

The major objective of the above SAR programme was to identify the structural features which underwrote the exceptional pharmacological profile demonstrated by prazosin both in vitro and in vivo, and then to apply this understanding to the design of improved analogues.  One approach focussed on the design of novel $\alpha_1$-adrenoceptor antagonists with improved duration of action over prazosin which would be suitable for once-daily administration in man to control elevated blood pressure.  Naturally, the 2,4-diamino-6,7-dimethoxyquinazoline nucleus was retained as a key building block and synthetic attention was focussed on elaboration of the 2-piperazino substituent.  Thus, replacement of the furan moiety of prazosin with a benzodioxan system, which was known to be compatible with $\alpha$-adrenoceptor blocking activity, provided doxazosin.[35]  This compound proved to be a potent, highly selective $\alpha_1$-adrenoceptor antagonist with long-lasting antihypertensive properties in rats and dogs.  In the latter species, 24hr control of blood pressure was clearly achieved after single daily doses (0.5mg/kg).  Heart rate was barely affected and there were no signs of tolerance after chronic dosing. Pharmacokinetic evaluation in dogs indicated an extended plasma half life when compared to prazosin[36] (4.7 vs 1.5 hr) and differences were even more apparent in man[37] (22 vs 2-3hr).  This marked improvement for doxazosin appears to derive from a lower plasma clearance rate presumably because the major route of metabolism for these quinazoline derivatives, 6/7-O-demethylation, is much less favoured.  Thus, it is interesting to note that structural modification in one area of such a complex molecule can have a profound influence on the molecular recognition processes which control acceptance of the distal methoxy functions by the O-demethylases.

Doxazosin has undergone extensive safety evaluation in animals with no untoward effects, and excellent toleration has also been observed in clinical studies to date.  As a result, doxazosin is receiving widespread approval by regulatory authorities for once-daily, first-line treatment of hypertension.  In addition to providing effective blood pressure control, doxazosin significantly reduced total cholesterol, LDL-cholesterol and triglycerides while significantly increasing the HDL-cholesterol to total cholesterol ratio.  The beneficial effects of doxazosin on blood pressure and lipid profile may favourably affect the risk of coronary heart disease.

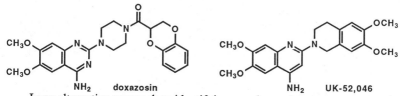

In an alternative approach to identifying novel $\alpha_1$-adrenoceptor antagonists with clinical utility, SARs in the 2,4-diaminoquinoline series (10) were examined in detail.[30] Although most derivatives displayed high $\alpha_1$-adrenoceptor activity, the binding affinity of UK-52,046 proved to be quite exceptional. Indeed, the $IC_{50}$ (6 x $10^{-12}$) is the lowest we have observed and represents a 30-fold improvement over prazosin. The receptor binding energy for UK-52,046 (-15.37 kcals/mole) is quite close to the Andrews value (-16.2 kcal/mole) and may reflect preference for a highly protonated (95%, pKa = 8.76), essentially coplanar system with limited degrees of freedom.

Pharmacological profiling of UK-52,046 in animals showed that the compound was effective in controlling cardiac arrhythmias provoked by adrenaline, ischaemia or reperfusion.[37,38] These observations suggest an important role for $\alpha_1$-adrenoceptors in the genesis of various types of arrhythmias. Preliminary studies in volunteers show that the $\alpha_1$-antagonist effects of UK-52,046 persist for up to 12h after a single intravenous dose (0.5 µg/kg) without marked effects on blood pressure or heart rate.[39] The potential for UK-52,046 to provide a novel mechanistic approach to the limited anti-arrhythmic therapies currently available, remains to be defined.

### Acknowledgements

Drug discovery and development is very much a team effort and the author is deeply indebted to the many individual scientists at Pfizer Central Research whose work is summarised in this chapter.

### References

1. Campbell, S.F., X-ray Crystallography and Drug Action, Horn, A.S., De Ranter, C.J. Eds; Clarendon: Oxford 1984; p347

2. Campbell, S.F. Second SCI-RSC Medicinal Chemistry Symposium; Emmett, J.C. Ed; Royal Society of Chemistry: 1984; p.18

3. Campbell, S.F., Davey, M.J., Hardstone, J.D., Lewis, B.N., Palmer, M.J. J. Med. Chem. 1987, 30, 49.

4. Alabaster, V.A., Campbell, S.F., Danilewicz, J.C., Greengrass, C.W., Plews, R.M. ibid, 1987, 30, 999

5.   Campbell, S.F., Plews, R.M. ibid, 1987, 30, 1794
6.   Campbell, S.F., Danilewicz, J.C., Greengrass, C.W., Plews, R.M. ibid, 1988, 31, 516
7.   Stanaszek, W.F., Kellerman, D., Brogden, R.N., Romankiewicz, J.A. Drugs, 1983, 25, 339
8.   For a recent review see: Br. J. Clin. Pharmacol., Reid, J.L., Davies, H.C. Eds. 1986, 21S
9.   For a recent review see: Am. Heart J., Hayduk, K. Ed. 1988, 116
10.  Starke, K., Montel, H., Gayk, W., Merker, R. Naunyn-Schmiedeberg's Arch. Pharmacol., 1974, 285, 133; Langer, S.Z. Pharmacol. Rev., 1981, 32, 337
11.  Cambridge, D., Davey, M.J., Massingham, R. Br. J. Pharmacol., 1977, 59, 514P
12.  For reviews see: Colucci, W.S. Am. J. Cardiol., 1983, 51, 639; Graham, R.M. ibid, 1984, 53, 16A
13.  $\alpha_1$- and $\alpha_2$-adrenoceptors in a rat brain membrane preparation were labelled with tritiated prazosin and clonidine respectively and the abilities of test compounds to displace these ligands measured[1]. Results are expressed as $K_i$ values (nM).
14.  Andrews, P.R., Craik, D.J., Martin, J.L. J. Med. Chem., 1984, 27, 1648. In this approach, binding energies are calculated from individual functional group values derived from analysis of 200 drugs and enzyme inhibitors.
15.  Crippen, G.M. ibid, 1979, 22, 988
16.  Griffiths, D.V., Swetnam, S.P. J. Chem. Soc. Chem. Comm., 1981, 1224; Brown, D.J., Teitei, T. J. Chem. Soc., 1965, 755
17.  Aue, D.H., Webb, H.M., Bowers, M.T. J. Am. Chem. Soc. 1976, 98, 311
18.  Saethre, L.J., Carlson, T.A, Kaufman, J.J., Koski, W.S. Mol. Pharmacol., 1975, 11, 492
19.  Carlström, D., Bergin, R. Acta. Cryst. 1967, 23, 313
20.  Hearn, R.A., Freeman G.R., Bugg, C.E. J. Am. Chem. Soc. 1973, 95, 7150.
21.  Zaagsma, J. J. Med. Chem., 1979, 22, 441
22.  Matthews, D.A., Bolin, J.T., Burridge, J.M., Filman, D.J., Volz, K.W., Kaufman, B.T., Beddell, C.R., Champness, J.N., Stammers, D.K., Kraut, J. J. Biol. Chem., 1985, 260, 381.
23.  Appebury, M.L., Hargrave, P.A. Vision Res. 1987, 26, 1881
24.  Cotecchia, S., Schwinn, D.A., Randall, R.R., Lefkowitz, R.J., Caron, M.G., Kobilka, B.K. Proc. Natl. Acad. Sci., 1988, 85, 7159.
25.  The piperidino derivative 8 was chosen as a convenient steric equivalent for most of the quinazolines listed in Table 3.
26.  Raffa, R.B., Porreca, F. Life Sciences, 1989, 44, 245.
27.  Weiland, G.A., Minnemann, K.P., Molinoff, P.B. Mol. Pharmacol., 1980, 18, 341

28. For a review of drug-receptor interactions see: Kollman, P.A., in Burger's Medicinal Chemistry, 4th Ed., Part 1, Wolff, M.E. Ed., Wiley, New York, 1980, p313

29. Carpy, A., Leger, J.M., Leclerc, G., Decker, N., Rouot, B., Wermuth, C.G. Mol. Pharmacol., 1982, 21, 400

30. Campbell, S.F., Hardstone, J.D., Palmer, M.J. J. Med. Chem., 1988, 31, 103

31. Bordner, J., Campbell, S.F., Palmer, M.J., Tute, M.S. ibid, 1988, 31, 1036

32. Howell, E.E., Villafranca, J.E., Warren, M.S., Oatley, S.J., Kraut, J. Science (Washington D.C.) 1986, 231, 1123

33. CLOG P3, Medicinal Chemistry Project, Pomona College, Claremont, California

34. Antihypertensive activity was evaluated after oral administration (5mg/kg) to spontaneously hypertensive rats (New Zealand or Okamoto strain). Falls in blood pressure (mm Hg) during the test period (4.5hr) were measured using an indirect tail cuff method and the maximum value was expressed as follows:-

$$\% \text{ reduction in hypertension} = \frac{\text{fall in blood pressure}}{\text{control blood pressure} - 130} \times 100$$

35. Campbell, S.F., Davey, M.J. Drug Design and Delivery, 1986, 1, 83

36. Kaye, B., Cussans, N.J., Faulkner, J.K., Stopher, D.A., Reid, J.L. Br. J. Clin. Pharmac. 1986, 21. 19S.

37. Conrad, K.A., Fagan, T.C., Mackie, M.J., Mayshar, P.V., Lee, S., Souhrada, J.F., Falkner, F.C., Lazar, J.C. Eur. J. Clin. Pharmacol., 1988, 35, 21.

38. Uprichard, A.G.C., Harron, D.W.G., Wilson, R., Shanks, R.G. Br. J. Pharmacol. 1988, 95, 1241

39. Flores, N.A., Sheridan, D.J. ibid 1989, 96, 670

40. Schäfers, R.F., Elliott, H.L., Howie, C.A., Reid, J.L. Br. J. Clin. Pharmacol. 1989, 27. 102P

# Molecular Recognition by Antibiotics of the Vancomycin Group. Dimerisation of the Antibiotics

By Dudley H. Williams and Jonathan P. Waltho

UNIVERSITY CHEMICAL LABORATORY, LENSFIELD ROAD, CAMBRIDGE CB2 1EW, UK

## 1 INTRODUCTION

Antibiotics of the vancomycin group have assumed increasing clinical importance during the last fifteen years, in part because of the increasing prevalence of Staph. aureus bacteria which are resistant to methicillin.[1] In addition, vancomycin itself has found extensive use in the treatment of post-operative diarrhoea, caused by Clostridium difficile in the gut. The antibiotic is then given orally, and has been found to be very efficient in curing a dangerous condition.

As the importance of the antibiotics in this group has increased, pharmaceutical companies in many parts of the world have initiated efforts to find new members. As a result, the group now consists of a large number of structures, all of which are hepta-peptides. For a vancomycin group antibiotic produced by any one actinomycete, there are often a number of variants, frequently differing in the nature of the attached sugars or fatty acid groups. The total number of structural variants which have been reported is in the region of one hundred (see, for example, Ref. 2).

For a number of years now, we have been engaged not only in the structure elucidation of members of this group, including the structure elucidation of vancomycin itself, but also in work designed to establish the molecular basis of their mode of action. The latter part of the work has been reliant on a finding[3] that vancomycin itself and another member of the group, ristocetin A (1), bind to cell wall mucopeptide precursors terminating in the dipeptide -D-Ala-D-Ala.

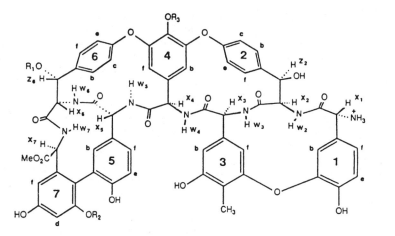

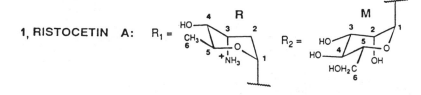

**1, RISTOCETIN A:**

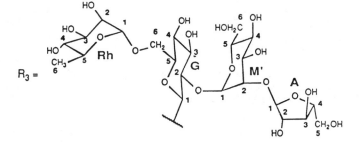

**2, RISTOCETIN Ψ: R₁= Ristosaminyl ; R₂=R₃=H**

Given this knowledge, we examined the proton magnetic
resonance spectra of ristocetin A in both the presence
and absence of the cell wall analogue
N-acetyl-D-Ala-D-Ala.[4]  Proton resonances of the
antibiotic in its free form, and when bound to the cell
wall analogue were assigned to specific protons in the
structure.  A similar analysis was carried out for the
proton resonances of the peptide.  With this information
in hand, we therefore knew the changes in chemical shift
of each proton resonance upon formation of a complex
between the antibiotic and the cell wall analogue.  In
particular, the changing chemical shift of NH-protons of
the antibiotic and cell wall analogue could be used to ·
indicate which of these protons, when they occur as part
of a secondary amide unit (-CO-NH-), are involved in
hydrogen bond formation.

This information can be accommodated by proposing
that in the complex between the antibiotic and the cell
wall analogue, the cell wall analogue is oriented
relative to the antibiotic as shown in Fig. 1.  In the
figure, dotted lines indicate hydrogen bond formation
between the carbonyl group of one component and an NH
group of the other component.  It can be seen that in
the proposed complex, the carboxyl group of the
C-terminal alanine of the cell wall analogue forms no
less than three hydrogen bonds to three NH groups which
lie in a pocket at one end of the antibiotic structure.
The hydrophobic walls of this pocket are created using
four of the phenolic sidechains of the constituent amino
acids of the antibiotic.  They are orientated such that
solvent molecules may be excluded from the polar groups
involved in the intermolecular hydrogen bonds of this
region.  As these interactions occur, an additional
hydrogen bond can be formed between the carbonyl oxygen
of the acetyl group of the cell wall analogue and an NH
which is seen at the left hand part of the antibiotic
structure.  The geometry of these hydrogen bonded
interactions allows simultaneously favourable
hydrophobic interactions to occur between the two
alanine methyl groups of the cell wall analogue and
portions of the benzene rings of the antibiotic.

The above model for the binding interaction has
been checked by use of the powerful technique of [4b]
intermolecular nuclear Overhauser effects (nOes).
This is a means of determining the spatial proximity of
hydrogen nuclei by observing their mutual relaxation
towards equilibrium from an excited state created within
the NMR experiment.  The nOe data for the
ristocetin-dipeptide complex showed that the binding

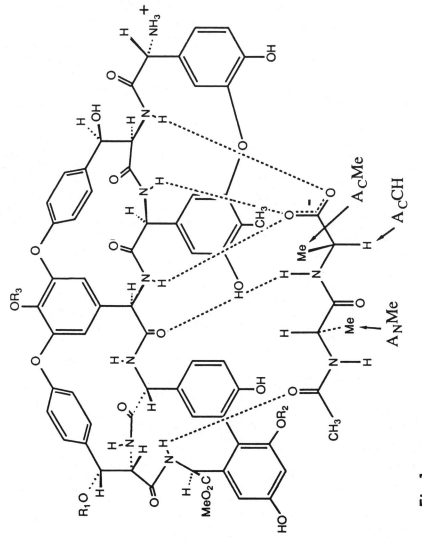

**Fig.1**– the ristocetin/cell wall analogue complex

model represented in Fig.1 is indeed correct in its
essential details. Those protons of the antibiotic and
cell wall analogue which are demanded to be near in
space by the binding model do give mutual nOes. Thus,
the molecular basis of action of the antibiotics is well
founded, and we have subsequently investigated how these
systems can be used to develop our understanding of
molecular recognition.

It is possible to measure the binding constant
between the two components by the use of UV-difference
spectroscopy.[5] These measurements have not only been
carried out for numerous antibiotics of the vancomycin
group with N-acetyl-D-Ala-D-Ala, but also to the
extended cell wall analogue,
di-N-acetyl-L-Lys-D-Ala-Ala. The binding constants in
general lie in the range $10^4$-$10^7$ 1 $mol^{-1}$, with the
somewhat larger binding constants (by about $10^1$-$10^2$)
normally being found for the more extended cell wall
analogue.

Given the above background information, in more
recent work we have been able to explore some of the
factors involved in defining the desired antibiotic
geometries, and factors involved in more subtle aspects[6]
of binding the cell wall analogues. This paper is[7]
concerned with the dimerisation of ristocetin.

## 2 DIMERISATION OF THE RISTOCETIN A/TRIPEPTIDE COMPLEX

It has been shown that certain resonances of[8]
ristocetin A and the ristocetin A-tripeptide complex[9]
in water are present at two different chemical shifts,
due to the presence of two forms in slow exchange. In
the former system, it was noticed that at room
temperature, two resonances of almost equal intensity
were present for the 6-methyl group of rhamnose ($\delta$, 0.98
and 1.28) - see 1. At 25°C, the interconversion rate[8]
of one form to the other was calculated to be ca 40 Hz.
Later studies[8] revealed that other proton resonances
were perturbed by this exchange process and identified
the involvement of proton 4b. In the latter system,
i.e. the ristocetin A-tripeptide complex, two sets of[6]
bound alanine methyl signals in slow exchange at 40°C
were observed. In a further study, carried out in 5:2[6]
$D_2O/CD_3CN$ solution at 22°C, from the presence of
saturation transfer crosspeaks between more than 2
resonances of certain protons (e.g. 2c, 4f, 6e, $G_1$,
$Rh_6$), it was apparent that more than two conformers were
present. For some protons, the resonances of two of the

forms had similar chemical shifts, whilst those of a
third form were shifted considerably (e.g. $x_4$ at 6.52.
6.55, 5.73). The resonances of 6e also occurred at
three distinct chemical shifts. Two resonances of 6e
(5.09 and 5.18) occurred close to those of 4f and so
must be experiencing <u>large</u> upfield ring current shifts,
whereas the resonance of the third form occurred as
expected for the monomer complex at 7.32 ppm. From this
chemical shift evidence, in conjunction with the
'unusual' nOes listed in Table 1, it became clear that
<u>intermolecular</u> processes (other than tripeptide binding)
were involved.

TABLE 1:  NOes observed within Ristocetin A (<u>1</u>) and
Ristocetin $\psi$ (<u>2</u>) in their tripeptide complexes in
$D_2O/CD_3CN$ solution that are inconsistent with the
covalent structure of the antibiotic (R=ristosaminyl
protons).

| | |
|---|---|
| $R_{2,2'}$ ↔ 2c | $R_3$ ↔ $X_3$ |
| $R_{2,2'}$ ↔ $X_3$ | 6f ↔ 4b |

     The nOes between protons of ristosamine (attached
to residue 6) and residues 2 and 3 were indicative of
the formation of a "head to tail" dimer through the
combination of the back faces of two ristocetin
molecules (with tripeptide still occupying the binding
site).  On inspection of molecular models, it was
noticed that there existed a high degree of
complementarity of hydrogen bond donors and acceptors
(as indicated in Fig. 2) as two molecules were placed
back to back, with $w_5$ opposite the carbonyl group of
residue 5, $(CO)_5$, and $w_6$ opposite $(CO)_3$.  Four
intermolecular hydrogen bonds are made in total
involving $w_5$ and $w_6$ from both molecules in the dimer
(see Figs. 2 and 3).  The large chemical shift changes
observed for the resonances of $x_4$ can be rationalised by
the proton being flanked by two $w_5$ to $(CO)_5$ hydrogen
bonds in two forms of the tripeptide complex but not in
the third (the monomer complex).  The conformation
enforced by the above hydrogen bonds places the proton
6e of each molecule (and to a lesser extent 6f) over the
face of ring 4 of the other molecule (see Fig. 4),
satisfying the observed upfield chemical shifts of their
resonances.  Indeed, all the 'unusual' nOes listed above
may be satisfied by the intermolecular proximity created
by a dimer of this nature.  This geometry also allows
indirect interaction of the protonated amine of

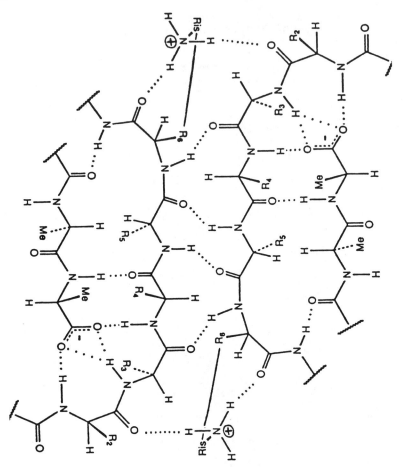

**Fig. 2** – the dimer complex

Fig. 3 - CPK model of dimer

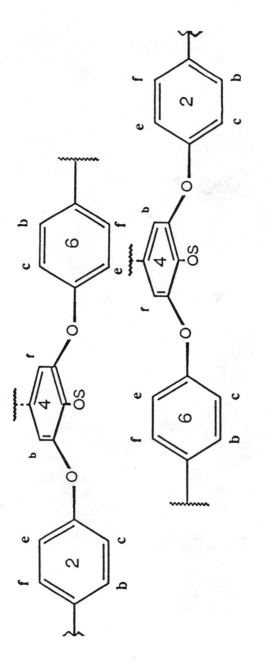

**Fig. 4** – relative orientation of aromatic rings in dimer

ristosamine of each molecule with the carboxylate anion
of the peptide bound to the other molecule through the
polarisation of the carbonyl group, $(CO)_2$, which is part
of an amide system involved in the carboxylate anion
binding pocket (see Figs. 2 and 3).

The ratio of the populations of the three forms
observed in the normal one-dimensional spectrum at room
temperature is approximately 5:5:1 using a 15 mM sample.
The low intensity form was shown to be the monomer by
diluting the sample by a factor of 10. The ratio of the
three forms was now approximately $1:1:4_1$ These ratios
indicate a binding constant of $2x10^3$ $M^{-1}$ for dimer
formation. From saturation transfer experiments, the
off-rate for the dimer was measured as $ca_4$ 8 Hz (and thus
dimerisation has an on-rate of ca $2x10^4$ $M^{-1}$).

The observation of two sets of dimer resonances may
be rationalised in terms of two different forms of the
dimer of equal energy in slow exchange, or asymmetry
within the dimer with a slow interconversion rate
between the two halves. It was not possible to
distinguish between the exchange rates of the two sets
of dimer resonances with the monomer using saturation
transfer experiments. It was also not possible to
distinguish between the intensities of the two sets of
dimer resonances. This is suggestive, though not
conclusive proof, that there exists a single, asymmetric
dimer.

Whether an asymmetric dimer or two different
symmetric dimers, the region of structural differences
in the dimer(s) may be probed by examining the protons
with the largest chemical shift differences of their
dimer resonances. The largest differences are observed
within the tetrasaccharide portion of the molecule, e.g.
$Rh_5$ $\Delta\delta$, 0.96 ppm; $G_1$ $\Delta\delta$, 0.60 ppm; $Rh_6$ $\Delta\delta$, 0.48 ppm,
with smaller changes for aglycone protons adjacent to
the tetrasaccharide, e.g. 2c $\Delta\delta$, 0.41 ppm; 6c $\Delta\delta$, 0.29
ppm; 2e $\Delta\delta$, 0.22 ppm. Thus, it appears that the
tetrasaccharide unit is differently orientated in the
dimer(s).

It is likely that one major difference between the
two forms is a result of the orientation of glucose with
respect to the aglycone portion, specifically with
respect to ring 4. The upfield glucose anomeric
resonance, and the corresponding $G_3$ and $G_5$ resonances,
share nOes with the downfield $A_cMe$ resonance indicating
that in this form the hydrophobic $G_1$, $G_3$, $G_5$ face of
glucose lies above the $A_cMe$ binding pocket (c.f. the

position of the hydrophobic face of vancosamine in
vancomycin complexes[10]). An nOe confirming this is
observed between $G_1$ (upfield) and 2e (downfield).

The corresponding nOes are not observed for the
other dimer conformer. It is likely that in this latter
form that the $G_1$, $H_3$, $G_5$ face of glucose lays on the
back side of the aglycone to which it is bonded, such
that $G_1$ lays over the edge of ring 6 of the dimer
partner. This would account for its relatively low
field resonance (5.55 ppm). The anomalously upfield
shifts of $Rh_5$ and $Rh_6$ in one form are indicative of it
being over the face of an aromatic ring. The nOe
between $Rh_6$ (upfield) and $R_1$ (downfield) is evidence that
the aromatic ring in question is ring 6 although it is
not known whether it is ring 6 of the same molecule or
the dimer partner. Furthermore, it is not known which
orientation of glucose with respect to ring 4 is
populated when the 5,6-region of rhamnose is over ring
6.

### 3 DIMERISATION OF THE RISTOCETIN $\psi$/TRIPEPTIDE COMPLEX

Inspection of the spectra of the ristocetin
tri-peptide complex reveals that a similar dimer/monomer
process is occurring in this system. The
inter-molecular nOes $R_1 \leftrightarrow 2c$, $R_2 \leftrightarrow 2c$ and $R_3 \leftrightarrow x_3$ are
observed, but the off-rate of the dimer is such that the
separate proton resonances are not observed for dimer
and monomer. The resonance most perturbed on dimer
formation in the ristocetin A-tripeptide complex, 6e( $\Delta\delta$
ca 2.2 ppm), is not observed in the ristocetin $\psi$
-tripeptide complex, presumably because its linewidth is
too large. Other strongly perturbed resonances in the
ristocetin A-tripeptide system, e.g. $x_4$ ( $\Delta\delta$, ca 0.8 ppm)
and 6f ( $\Delta\delta$, ca 0.7 ppm), are present in the ristocetin
-tripeptide spectra as single, <u>broad</u> resonances,
indicating that they are only just on the fast exchange
side of the coalescence point. Therefore, the off-rate
of the ristocetin $\psi$ -tripeptide dimer at room
temperature must be greater than ca 300 Hz, compared
with ca 8 Hz for the ristocetin A-tripeptide dimer. As
a sample of the ristocetin $\psi$-tripeptide complex is
heated, the resonance of, for example, $x_4$ sharpens <u>and</u>
moves upfield indicating that the equilibrium constant
for dimerisation falls with temperature. Thus, it has a
negative $\Delta S$ component. It may be concluded that it is
the tetrasaccharide attached to ring 4 that slows the
kinetics of dimer separation (the ring 7 mannose being
on the opposite face to the intermolecular interaction).

## 4 DIMERISATION OF OTHER SPECIES

The dimerisation phenomenon is probably not unique to ristocetin complexes, aqueous solvents or even ristocetin. For example, in the spectrum of ristocetin A in $D_2O-CD_3CN$ solution, the resonances of the two forms observed for the $Rh_6$ methyl group are of identical chemical shift to the two forms of the <u>dimer</u> in the ristocetin A-tripeptide complex; in the spectrum of ristocetin $\psi$ in $D_2O-CD_3CN$ solution, the resonances of $x_4$ and 6e are again very broad; in that of ristocetin $\psi$ in DMSO solution[11] the resonances of $x_3$ and 6f join those of $x_4$ and 6e in having large linewidths; in that of A40926-PS in DMSO solution[11] the resonances of residue 3 are particularly broad. Vancomycin appears to dimerise in aqueous solution but not in DMSO solution although interpretation of linewidth phenomena is slightly confused by a dynamic process involving the $w_3$ amide unit.[12] Indeed, in studies of the specific rotation of vancomycin in 0.02 M citrate buffer,[13] it was noted that significant aggregation of the antibiotic occurred at millimolar concentrations. At 10 mM, aggregation was judged to be total, and although no geometry of the aggregated species was determined, an association constant of $8 \times 10^2$ $M^{-1}$ was calculated, assuming dimerisation to have occurred. This figure is comparable with the association constant of $2 \times 10^3$ $M^{-1}$ calculated above for the ristocetin A-tripeptide complex. Dimerisation may also explain the two conformers recently reported for teicoplanin $A_2$.[14] The downfield shift of $x_4$ ($\Delta\delta$, 0.61 ppm) is similar to that observed for ristocetin but small changes for the resonance of 6e suggest that if a dimer is formed, its geometry is significantly different to that of ristocetin. The other resonances shifted considerably for teicoplanin are 3b ($\Delta\delta$, 0.23 ppm) and 5f ($\Delta\delta$, 0.23 ppm). The critical micelle concentration of teicoplanin (at pH 7.4) has been measured[15] to be 210 M suggesting that intermolecular interaction <u>is</u> prevalent in this system at the concentrations used in NMR studies.

## 5 DOES DIMERISATION HAVE A BIOLOGICAL ROLE?

It is possible that dimerisation has a role in the biological function of ristocetin. For example, the affinity of one ristocetin molecule for another might promote the siting of a second antibiotic molecule with one already bound to a cell-wall precursor terminating in -D-Ala-D-Ala. If such cell-wall precursor molecules are concentrated locally, then the antibiotic could thereby be targetted more effectively. Alternatively,

the same end might be achieved by delivery of antibiotic
dimers to locally concentrated cell-wall precursors
terminating in  -D-Ala-D-Ala.  However, at the minimum
inhibitory concentrations of ristocetin (approximately $5\mu$
M)[16], this latter possibility seems unlikely since the
antibiotic should occur mainly as the monomer in
solution.

   There are indications that dimerisation may be most
strongly promoted in aqueous media, e.g. ristocetin A
dimerises in $D_2O$, but much less so (if at all) in DMSO.[8]
If this is generally true, then the fact that the
off-rate for the dimer of ristocetin A in $D_2O$ is 40 Hz,
whereas that for the dimer of the ristocetin
A/tripeptide complex in 5:2 $D_2O/CD_3CN$ is 8 Hz, implies
an even lower value for the latter if this rate constant
were determinable in $D_2O$.  The significantly larger
barrier to the off-process for the dimer of the complex
(relative to that for the simple dimer) implies that
dimerisation is promoted by filling the antibiotic
binding cleft with the cell wall analogues.  This
implication is physically reasonable since it has
already been noted that the carboxylate anion of the
cell wall analogue (bound to one half of the antibiotic
dimer) can indirectly interact with the protonated amine
of ristosamine (in the other half of the antibiotic
dimer).  Additionally, the "back-faces" of ristocetin A
which are involved in the dimerisation (Fig. 2) should
have a greater mutual affinity the more remote are the
hydrogen bonds formed between the faces from bulk water.
That is, these hydrogen bonds are more buried from bulk
water, and thereby strengthened, when the cell wall
analogues are bound.  In a similar way, the hydrogen
bonds formed deep in the interior of proteins should in
general be stronger than those near the surface.  In
this sense, the aggregates discussed in this paper model
some aspects of protein folding.

### 6 CONCLUSION

   The different forms observed in a ristocetin
A-tripeptide complex in $D_2O$-acetonitrile-$d_6$ solution are
the result of the formation of either an asymmetric
dimer, or two symmetric dimers, with an association
constant of ca $2x10^3$ $1M^{-1}$.  The former explanation is
the more probable. The geometry of the two forms of the
dimer, as determined by chemical shift changes and
intermolecular nOes, is such that the back sides of two
molecules come together forming hydrogen bonds along the
antibiotic amide backbone.  This causes ring 6 of each
molecule to lie close to the face of ring 4 of the

other.  The difference between the two forms of the
dimer appears to be a result of the orientation of the
tetrasaccharide attached to ring 4 of ristocetin.
Primarily, it appears that the hydrophobic $G_1$, $G_3$, $G_5$
face of glucose occupies one side of ring 4 or the
other.  Also, in one form, the $Rh_5$, $Rh_6$ region lays over
the face of ring 6.

Similar dimerisation is observed in the ristocetin $\psi$
-tripeptide complex although the off-rate of the dimer
is considerably faster.  This further implicates the
tetrasaccharide in a specific role in dimer formation in
addition to causing asymmetry.  It appears that
dimerisation also occurs, but to a lesser extent, in
DMSO solution, and may be responsible for some of the NMR
phenomena observed for vancomycin, A40926 and
teicoplanin $A_2$.  The geometry of the dimer is such that
the intermolecular hydrogen-bonds are shielded from the
solvent by the coming together of large hydrophobic
regions of the amino acid sidechains within the
antibiotic.  This is analogous to the binding of the
target peptide on the opposite face of the antibiotic.

ACKNOWLEDGEMENTS

Financial support from the Upjohn Company, Smith
Kline and French Research, U.K., and the SERC is
gratefully acknowledged.

REFERENCES

1.  R. Wise and D. Reeves (Eds.), J. Antimicrob.
    Chemother., 1984, 14, Suppl. D.

2.  J.C.J. Barna and D.H. Williams, Ann. Rev.
    Microbiol., 1984, 38, 339.

3.  H.R. Perkins, Biochem. J., 1969, 111, 195.

4.  (a) J.R. Kalman and D.H. Williams, J. Am. Chem.
    Soc., 1980, 102, 897;  (b) J.R. Kalman and D.H.
    Williams, J. Am. Chem. Soc., 1980, 102, 906.

5.  M.Nieto and H. R. Perkins, Biochem. J., 1971, 123,
    773.

6.  D.H. Williams and J.P. Waltho, Biochem.
    Pharmacology, 1988, 37, 133.

7.  J.P. Waltho and D.H. Williams, J. Am. Chem. Soc., in
    press.

8.  D.H. Williams, V. Rajananda, and J.R. Kalman, J.
    Chem. Soc., Perkin Trans. I, 1979, 787.

9.  M.P. Williamson and D.H. Williams, J. Chem. Soc.
    Perkin Trans. I, 1985, 949.

10. K. Rajamoorthi, C.M. Harris, T.M. Harris, J.P.
    Waltho, N.J. Skelton and D.H. Williams, J. Am. Chem.
    Soc., 1988, 110, 2946.

11. J.P. Waltho and D.H. Williams, unpublished results.

12. J.P. Waltho, D.H. Williams, D.J.M. Stone, N.J.
    Skelton, J. Am. Chem. Soc., 1988, 110, 5638.

13. M. Nieto and H.R. Perkins, Biochem. J., 1971, 123,
    773.

14. S.L. Heald, L. Mueller, and P.W. Jeffs, J. Mag.
    Res., 1987, 72, 120.

15. A. Corti, A. Soffientini and G. Cassani, J. Appl.
    Biochem., 1985, 7, 133.

16. R.V. Nielsen, F. Hyldig-Nielsen and J. Jacobsen, J.
    Antibiotics, 1982, 35, 1561.

# NMR Approaches to Recognition: Dihydrofolate Reductase

## By G. C. K. Roberts

DEPARTMENT OF BIOCHEMISTRY AND BIOLOGICAL NMR CENTRE, UNIVERSITY OF LEICESTER, LEICESTER LE1 7RH, UK

## 1 INTRODUCTION

Recent developments in molecular and structural biology have transformed our ability to explore the structure-function relationships of biological macro-molecules and, in particular, to obtain new insights into the molecular recognition processes which form the basis of biological specificity. It is now beginning to become possible to make quantitative estimates of the thermodynamic, structural and dynamic contributions of individual interactions to the overall molecular recog-nition process. This requires detailed functional and structural (spectroscopic and crystallographic) studies of the molecular interactions, coupled with precisely defined structural changes in each partner in the interaction, made by genetic or chemical means.

Dihydrofolate reductase (dhfr) is the 'target' for the important 'anti-folate' drugs such as methotrexate, trimethoprim and pyrimethamine. Its pharmacological importance, coupled with its conveniently small size ($M_r$ 18-25,000 in most species), has led to intensive studies of this enzyme over the last 10-12 years (1-4), including detailed structural studies by both crystall-ography and nmr. This considerable background of information allows us to ask precise questions about the roles of individual amino-acid residues in substrate, inhibitor and coenzyme binding. At the same time, the tools are available to allow us to provide quantitative answers to these questions; nmr, mutagenesis and kinetic studies of the protein are all well-developed.

The information and methodology available make dhfr well suited as a 'test bed' for studies of the details of molecular recognition in one particular system. In this paper, some of our recent studies of substrate, inhibitor and coenzyme binding to L. casei dhfr will be described, with particular emphasis on the use of nmr spectroscopy.

## 2  STRUCTURAL & FUNCTIONAL COMPARISONS

The effects of changing the structure of both partners on the protein-ligand interactions can readily be measured in this system. An enormous range of dhfr inhibitors has been synthesised over the last 40 years (1), while much more recently reliable protocols for making specific changes in the protein structure by oligonucleotide-directed mutagenesis have been developed (e,g., for dhfr, 2), and fluorimetric methods for equilibrium and kinetic measurements of ligand binding have been developed (e.g., 3, 4). If the changes in strength of binding with changes in chemical structure of ligand or protein are to be reliably interpreted, the three-dimensional structures must also be compared in detail. High-resolution crystal structures have been determined for dhfr from L.casei and other organisms in a number of ligand complexes (5-9), and crystal structures are also available for mutants of the E. coli and human enzymes (10, 11). In the case of L. casei dhfr, we have used nmr to make structural comparisons between complexes with different ligands (12-14) and between wild-type and mutant enzyme (14-16). [1]H resonances from more than 25% of the residues of L. casei dhfr have been assigned (14, 17, 18), and as a result 'reporter groups' are available throughout the protein structure; in addition, [1]H, [13]C, [15]N, [19]F and [31]P resonances of bound ligands can be used to characterise and compare their modes of binding (e.g., 13-16, 19-23).

A number of comparisons between related ligands, notably the important one between between the inhibitor methotrexate and the substrate folate, are discussed below. In addition, nmr can be used to compare the binding of different conformational isomers of the same molecule. For example, we have studied the binding of a series of pyrimethamine analogues synthesised by Stevens and his colleagues at Aston. Rotation about the phenyl-pyrimidine bond of pyrimethamine is, as expected, slow, and in unsymmetrically substituted compounds such as

2,4-diamino- 5-(4-fluoro, 3-nitrophenyl)- 6-ethyl pyrimidine we have found that the two conformational isomers
of the inhibitor (related by a rotation of approx. 180°
about the phenyl-pyrimidine bond) both bind to the
enzyme, with similar but not equal affinities (Figure 1;
22).

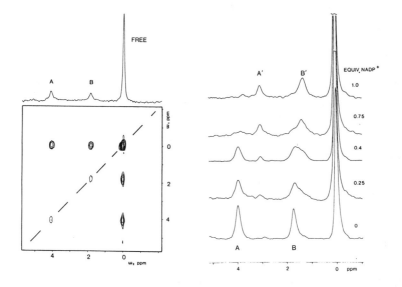

**Figure 1.** $^{19}$F nmr studies of the binding of 2,4-
diamino-5-(4-fluoro, 3-nitrophenyl)-6-ethylpyrimidine to
L. casei dihydrofolate reductase. **Left,** binary complex;
the top spectrum shows the resonances of free ligand and
of the two conformers, A and B, of the ligand bound to
the enzyme. The two conformers are related by a rotation
of approx. 180° about the phenyl-pyrimidine bond. Below
is a 2D NOESY/exchange spectrum, demonstrating that the
two species A and B exchange readily with the free
ligand but not with each other (as expected from the
severely restricted rotation in o-substituted biphenyl-
like systems). **Right,** the ternary complex, showing the
changes in chemical and in relative intensity of the two
resonances of the bound ligand on addition of NADP$^+$ to
form the ternary complex. NADP$^+$ binding alters the
relative affinity of the enzyme for the two conformers.
All spectra are referenced to the signal of the free
ligand. From ref. 22 with permission.

Comparison of the structures of mutant and wild-type protein both by X-ray crystallography and by nmr has revealed considerable variation in the structural consequences of amino-acid substitutions. Crystallographic studies of the Asp27 -> Asn mutant of E. coli dhfr (10) and nmr studies of the corresponding L. casei mutant, Asp26 -> Asn (24), show that any structural changes in the enzyme-methotrexate complex are very small and local (and may result from the displacement of a bound water molecule). In dhfr Trp21 -> Leu, the majority of the assigned resonances are again unaffected. However, in this case small chemical shift differences are seen for resonances of seven residues which are not close to the site of substitution - some as much as 10 A away (15). Finally, the Thr63 -> Gln substitution involves a residue which hydrogen-bonds to the 2'-phosphate of the coenzyme. Nmr analysis shows that the conformational effects are again slight, but, surprisingly, they are concentrated in the active site and not around the adenine binding site (25). This is thus a striking example of the effects of a mutation being transmitted a substantial distance through a protein. A possible mechanism for the transmission of these effects from the site of substitution is by a slight movement of helix C.

Up to now, the structural comparisons of mutant dhfrs and of complexes with different ligands have largely been made on the basis of chemical shift comparisons. These provide a simple and sensitive method for the detection of structural differences, but it is not generally possible to use observed shift differences to describe the precise nature of the difference in conformation. Our preliminary results suggest that nuclear Overhauser effects (NOEs) can be quantitated with sufficient accuracy to permit a detailed description of these (generally) small structural differences, and work is in progress to develop the necessary methods.

## Kinetic studies

In collaboration with Dr. C. Fierke and Prof. S. Benkovic (Pennsylvania State University) we have undertaken detailed studies of the kinetic mechanism of L. casei dhfr (26). The most striking feature of this mechanism is the kinetically preferred pathway for product release, in which tetrahydrofolate ($FH_4$) dissociates only after NADPH has bound. This originates from the marked negative cooperativity in binding between $FH_4$ and NADPH; $FH_4$ dissociates 300 times faster from the

E.NADPH.$FH_4$ complex than from E.$FH_4$. Very similar negative cooperativity between NADPH and the stable product analogue 5-formyl-$FH_4$ was earlier demonstrated by nmr and binding constant measurements (27). An interesting consequence of this mechanism is that the free enzyme is not on the kinetically preferred pathway.

At pH 6.5, the major contribution to the rate-limiting step is the dissociation of $FH_4$ from the E.NADPH.$FH_4$ complex, and there is thus only a small $^2H$ isotope effect with $NADP^2H$; this isotope effect, and hence the contribution of hydride transfer to the rate-limiting step, increases with increasing pH as $k_{cat}$ decreases Dhfr Trp21 -> Leu (W21L) has a $k_{cat}$ value 40-fold lower than that of the wild-type at pH 6.5. However, there is an increased $NADP^2H$ isotope effect on $k_{cat}$ in dhfr W21L, indicating a change in the contributions to the rate-limiting step. Detailed kinetic studies (26) reveal that the rate of $FH_4$ dissociation is unaffected in dhfr W21L, but the rate of hydride transfer has decreased by a factor of 100. This illustrates the importance of a detailed kinetic analysis of the mutant enzymes.

### 3  ELECTROSTATIC INTERACTIONS AND HYDROGEN BONDING

The binding of the inhibitor methotrexate involves a number of important ion-pairs and a network of hydrogen bonds between enzyme and ligand which must contribute to binding and specificity. We have studied the His28-Y-carboxylate and Arg57-o-carboxylate interactions using methotrexate analogues (12, 13). The Y-amide of methotrexate binds 9 times less tightly to the enzyme than methotrexate itself. Of the assigned resonances, only those of His28 and Leu27 differ between the Y-amide and methotrexate complexes; indicating that these complexes are essentially isostructural, and that the difference in binding constant is a reasonable measure of the contribution of the His28-Y-carboxylate ion pair to the overall interaction energy. The o-amide of methotrexate binds 100-fold less tightly than methotrexate, but the differences in chemical shifts of the assigned resonances indicate that in this case the whole of the p-amino-benzoyl-glutamate moiety binds differently in the analogue, so that the difference in binding energy cannot be simply interpreted (12, 13).

An electrostatic interaction which is particularly important in inhibitor binding is that between Asp26

(Asp27 in <u>E. coli</u> dhfr) and the protonated pteridine
ring of methotrexate (28) or the corresponding protonated pyrimidine ring of trimethoprim (29). By contrast,
the pteridine ring of the substrate, folate, which binds
in a different orientation (see below), binds in the
uncharged state (30). For both inhibitors, the pK of $N_1$
is markedly increased on binding (28, 29), showing that
the charged form binds considerbly more tightly (as much
as $10^5$-fold) than the neutral form. The Asp27 -> Asn
mutant of the <u>E. coli</u> enzyme binds methotrexate only 27-
fold less tightly than the wild-type at neutral pH (10),
but it is difficult to use this comparison to assess the
contribution of the ion-pair, since methotrexate is in
the <u>neutral</u> (high pH) state in the complex (10), but the
asparagine residue 'mimics' the <u>protonated</u> (low pH) form
of aspartate.

These electrostatic interactions also involve
hydrogen bonds, and nmr offers an excellent means of
monitoring hydrogen bonds between ligand and protein.
For example, the carboxylate of Asp26 accepts hydrogen
bonds from $N_1$-H and the 2-$NH_2$ of the bound inhibitor,
while the second proton of the 2-amino group of
methotrexate (and, probably, folate) hydrogen bonds
through a water molecule to Thr116. By using specifically $^{15}$N-labelled trimethoprim (31) and methotrexate
(32), together with inverse detection experiments, we
have located the resonances of the protons involved in
all three of these hydrogen bonds, allowing us to
monitor them very directly.


### 4   NON-POLAR CONTACTS

The packing of non-polar residues around the
ligands, notably around the pteridine and benzoyl rings
of methotrexate and folate, makes an important contribution to binding. Among the residues making hydrophobic
contacts with methotrexate are Leu4, Leu19, Leu27,
Phe30, Phe49 and Leu54. Resonances of all these residues
have been assigned, and their packing around the
inhibitor is well defined by a network of NOEs. These
NOEs, together with chemical shift changes provide
sensitive measures of changes in packing produced by
mutagenesis or by changes in inhibitor structure. For
example, comparison of the binding of 3',5'-difluoro-
and 3',5'-dichloro-methotrexate to that of methotrexate
itself shows that the orientation of the benzoyl ring
about its symmetry axis in the complex changes by up to
$25^\circ$ to accomodate the bulky substituents (13).

The effects of altering non-polar contacts by site-directed mutagenesis can be illustrated by dhfr Trp21 -> Leu. The major effect of this substitution in terms of binding is on the coenzyme; the binding constant of NADPH is decreased 400-fold but that of NADP$^+$ only 2-fold (15). The indole ring of Trp 21 makes contact with the amide substituent on the nicotinamide ring of the bound coenzyme, but a leucine side-chain at this position is too small to do so. This difference appears to lead to a change in the orientation of the nicotinamide ring in the site, as indicated by differences in chemical shift for the protons of this ring in the complex (15), and hence to decreases in $k_{cat}$ and binding constant. The marked difference in the effect of the substitution on the binding of NADPH and NADP$^+$ may be explained by our observation (e.g. 3) that the oxidised nicotinamide ring makes little contribution to the binding of NADP$^+$. There is also an interesting parallel between the effects of the W21L substitution and those of replacing the nicotinamide carboxamide by a thioamide to give thioNADPH. In both cases there is a marked decrease in $K_a$ and $k_{cat}$, and an apparent change in the mode of binding of the nicotinamide ring; these effects probably arise in <u>both</u> cases from a perturbation of the interaction between Trp 21 and the nicotinamide.

## 5   STRUCTURAL DYNAMICS

Like all molecules, proteins and their complexes with ligands have structures which fluctuate at a wide variety of rates; since molecular recognition is, by its nature, a dynamic process, it is important to characterise these fluctuations as well as the time-average structure. Nmr has permitted the identification and partial characterisation of structural fluctuations in dhfr at rates ranging from 1 s$^{-1}$ to $10^9$ s$^{-1}$. Slow conformational equilibria (see below) have been observed in the enzyme-trimethoprim-NADP$^+$ (33, 34), enzyme-folate-NADP$^+$ (35-37) and enzyme-folinic acid (27) complexes. Fluctuations in the conformations of bound ligands over a wide range of amplitudes and rates have been characterised for 3',5'-difluoromethotrexate (21) and trimethoprim (38).

Lineshape analysis shows that the benzoyl ring of 3',5'-difluoromethotrexate rotates (or rather 'flips') about its symmetry axis by 180$^{\circ}$ at rates of approximately $10^3$ s$^{-1}$ at room temperature (21). This rate is determined, in large measure, by the packing of

amino-acid side-chains around the ring, and their
ability to move so as to allow the ring to flip, and it
thus provides a sensitive measure of changes in non-
polar interactions around the ring. For example, the
binding of the coenzyme NADPH increases the affinity of
the enzyme for methotrexate, at least in part through
ligand-induced conformational changes. One of these
involves a movement of helix C, which runs from the
adenine binding site to the benzoyl ring binding site
(17); the C-terminal residue of this helix is Phe49,
which is in contact with the benzoyl ring, and the
changes in packing around the ring resulting from the
coenzyme-induced conformational change are reflected in
a 3-fold increase in the rate of ring flipping.

Similar studies of the symmetrically substituted
benzyl ring of trimethoprim have been made using [m-
methoxy $^{13}$C]-trimethoprim (38). Relaxation measurements
revealed fluctuations of $\pm$ 25-35$^{\circ}$ at rates of about 10$^9$
s$^{-1}$, while lineshape analysis shows that 180$^{\circ}$ 'flips'
about the symmetry axis occur at 250 s$^{-1}$ at room
temperature. In trimethoprim, a major contribution to
the barrier to this 'flipping' comes from steric interactions within the trimethoprim molecule, between the
benzyl and pyrimidine rings, and molecular mechanics
calculations show that the 'flipping' of the benzyl ring
must be accompanied by rotation of about 60$^{\circ}$ about the
C5-C7 bond so as to change the relative orientation of
the two rings in bound trimethoprim (38).

The use of $^{15}$N-labelled trimethoprim to identify the
resonance from the proton in the $N_1$H-Asp26 hydrogen bond
has been mentioned above; lineshape analysis of this
resonance yields an estimate of the rate of exchange of
this proton with the solvent, and hence of the rate of
making and breaking of this crucial hydrogen-bond
between inhibitor and enzyme; this occurs at a rate of
34 s$^{-1}$ at room temperature (38), and clearly results
from a fluctuation in the protein structure distinct
from that involved in the ring flipping. The rates of
all these fluctuations in the structure of the enzyme-
trimethoprim complex are affected by coenzyme binding
(38), and also by amino-acid substitutions. Dhfr Asp26
-> Glu has kinetic parameters closely similar to those
of the wild-type enzyme (26); nmr studies (15) corres-
pondingly indicate close structural similarity, but with
some evidence for slight movements of the pteridine ring
and of helix B to accommodate the larger side-chain.
This structural similarity is accompanied by clear
differences in the dynamics of bound trimethoprim. The

rates of the conformational fluctuations discussed above
are increased by as much as 30-fold in this mutant (15).
It appears that structural fluctuations are significantly more sensitive to the effects of amino-acid
substitutions than is the time-average structure.

## 6 CONFORMATIONAL EQUILIBRIA

As noted above, there is good evidence for
conformation changes accompanying ligand binding which
contribute to the specificity of the enzyme (e.g., the
postulated movement of helix C on coenzyme binding; 17).
In addition, there is evidence for conformational
equilibria in the complexes which have been partially
characterised by nmr.

In spite of the close structural similarity
between methotrexate and folate, the stereochemistry of
reduction implies a difference of $180^\circ$ in the orientation of the pteridine ring between substrate and
inhibitor (39-41). We have demonstrated that both the
enzyme-folate and the enzyme-folate-NADP$^+$ complexes
exist in solution as a mixture of two or three slowly
interconverting conformations whose proportions are pH-
dependent (35-37). Comparison of chemical shifts between
the three conformations of the dhfr-folate-NADP$^+$ complex
indicate that the structural differences are localised
to the active site region, close to the pteridine and
nicotinamide rings. 2D NOE experiments (Figure 2) have
recently shown that one crucial difference between these
conformations is the orientation of the pteridine ring.
In the conformations denoted I and IIa, NOEs are
observed between the pteridine 7-proton of bound folate
and the two methyl groups of Leu27, indicating that the
ring has the same orientation in the binding site as
that of methotrexate. In conformation IIb, by contrast,
no such NOEs are observed; given the structural
constraints of the pteridine ring binding site, this can
only be explained if the ring has turned over by
approximately $180^\circ$ about an axis along the C2-NH$_2$ bond
(37). Only the latter is a catalytically productive mode
of binding for the substrate. In this case, therefore,
the substrate is able to bind, with very similar
affinity, in both productive and non-productive orientations; the inhibitor methotrexate binds only in the
non-productive orientation, in which it can form the
ion-pair with Asp26 discussed above.

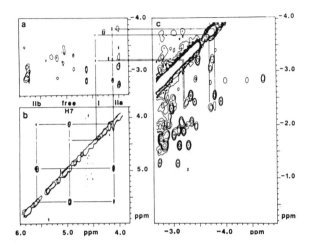

**Figure 2.** Nuclear Overhauser effects involving the folate pteridine 7-proton and methyl groups of the protein in the three conformations of the dhfr-folate-NADP$^+$ complex. Part b shows the aromatic region of the 2D exchange spectrum obtained from a sample containing the complex and a 2-fold excess of free folate. The cross-peaks linking the 7-proton of free folate at 4.95 ppm to the resonances of the same proton in the three conformations of the complex are connected by lines. Part a shows the region of a 2D NOESY spectrum containing cross-peaks between aromatic and methyl proton resonances. Cross-peaks involving the 7-proton resonance in conformations I and IIa are indicated; there are no cross-peaks in this region of the spectrum involving the 7-proton resonance in conformation IIb. Part c shows the high-filed region of the COSY spectrum of the complex. The lines connect the resonances of the 7-proton in conformations I and IIa (in part b) with their NOESY cross-peaks in part a and with the corresponding COSY cross-peaks in part c. The latter help to identify the methyl protons giving NOEs to the 7-proton as those of Leu27. The spectra are referenced to internal dioxan. From ref. 37 with permission.

------------------------------------------------------------

The complex of folate and NADP$^+$ with dhfr Asp26 -> Asn exists in only a single conformation over the pH range 5.5-7.5; this is demonstrated, for example, by the behaviour of the methyl resonances of Leu118 shown in Figure 3. The single conformation seen for this mutant corresponds to the low-pH conformer of the wild-type complex (16), strongly suggesting that Asp26 is the group responsible for the pH-dependence of the conformational equilibrium.

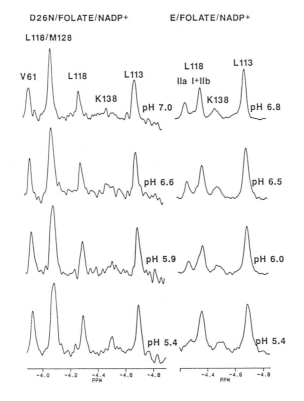

**Figure 3.** Comparison of the pH-dependence of the high-field region of the $^1$H spectra of the complex of folate and NADP$^+$ with, **right**, wild-type dhfr and, **left**, the Asp27 -> Asn mutant. In the wild-type complex, the methyl resonance of Leu118 has two components at high pH, corresponding to conformations IIa and (I + IIb) respectively; in the mutant complex, however, only a single, pH-independent resonance from Leu118 is observed. From ref. 16.

The complex between dhfr, trimethoprim and NADP$^+$ also exists in a conformational equilibrium; this is independent of pH, and involves primarily the nicotinamide ring of the coenzyme. In one conformation, this ring is bound specifically to a site on the enzyme, but in the other, rotations about the nicotinamide ribose C5'-O and pyrophosphate P-O bonds have altered the conformation of the coenzyme so that the nicotinamide ring is no longer in contact with the enzyme, but hanging free in solution, the coenzyme being bound solely through its adenosine moiety (33, 34). The fact that these two conformations are almost equally populated shows that the oxidised nicotinamide ring contributes very little to the overall binding energy. The relative populations of the two conformations are readily altered by changes in coenzyme or inhibitor structure (34) or by amino-acid substitutions in the protein (15).

## 7  CONCLUSIONS

The tools presently available for the study of molecular recognition, from X-ray crystallography and nmr spectoscopy to site-directed mutagenesis, have reached unprecedented levels of sophistication, allowing us to ask very precise questions about the details of the recognition process. Nmr spectroscopy contributes to this in three main ways: by allowing one to monitor individual protein-ligand interactions, such as hydrogen bonds; by providing a convenient and sensitive means for comparing the structure of wild-type and mutant proteins in solution; and by permitting the characterisation of dynamic processes in proteins and their complexes over a wide range of rates. The limitations of the technique, notably in the size of protein that can be tackled, are gradually becoming less restrictive, notably with the continuing development of higher-field spectrometers and the use of $^{13}$C and $^{15}$N labelling in conjunction with heteronuclear multiple quantum coherence experiments. Nmr is now an essential tool in the armoury of the protein scientist.

Notwithstanding the dramatic technical advances over the last few years, both in nmr and in other areas, a truly quantitative understanding of molecular recognition remains elusive, even in a system as well-characterised as dihydrofolate reductase. The conformational flexibility of proteins, and the subtlety of their responses to amino-acid substitutions and changes in ligand structure, continue to present substantial challenges to our experimental ingenuity.

## 8 ACKNOWLEDGEMENTS

The work described here is part of a continuing collaboration with Jim Feeney and Berry Birdsall (Mill Hill) and Julie Andrews (Manchester). Many people, whose names appear in the reference list, have made essential contributions to the experiments and their interpretation; my thanks to all of them. Work at Leicester has been supported by SERC (Protein Engineering Club) and the Wellcome Trust.

## 9 REFERENCES

(1) B. Roth and C. C. Cheng, <u>Progress in Medicinal Chemistry</u>, 1982, <u>19</u>, 270.
(2) J. Andrews, P. F. G. Sims, S. Minter & R. W. Davies, submitted for publication
(3) B. Birdsall, A. S. V. Burgen and G. C. K. Roberts, <u>Biochemistry</u>, 1980, <u>19</u>, 3723.
(4) S. M. J. Dunn, J. G. Batchelor and R. W. King, <u>Biochemistry</u>, 1978, <u>17</u>, 2356.
(5) J. T. Bolin, D. J. Filman, D. A. Matthews, R. C. Hamlin, and J. Kraut, <u>J. Biol. Chem.</u>, 1982, <u>257</u>, 13650.
(6) D. J. Filman, J. T. Bolin, D. A. Matthews and J. Kraut, <u>J. Biol. Chem.</u>, 1982, <u>257</u>, 13663.
(7) K. W. Volz, D. A. Matthews, R. A. Alden, S. T. Freer, C. T. Hansch, B. T. Kaufman and J. Kraut, <u>J. Biol. Chem.</u>, 1982, <u>257</u>, 2528.
(8) D. A. Matthews, J. T. Bolin, J. M. Burridge, D. J. Filman, K. W. Volz, B. T. Kaufman, C. R. Beddell, J. N. Champness, D. K. Stammers, and J. Kraut, <u>J. Biol. Chem.</u>, 1985, <u>260</u>, 381.
(9) C. Oefner, A. D'Arcy and F. K. Winkler, <u>Eur. J. Biochem.</u>, 1988, <u>174</u>, 377.
(10) E. E. Howell, J. E. Villafranca, M. S. Warren, S. J. Oatley, and J. Kraut, <u>Science</u>, 1986, <u>231</u>, 1123.
(11) A. J. Geddes, personal communication.
(12) D. J. Antonjuk, B. Birdsall, H. T. A. Cheung, G. M. Clore, J. Feeney, A. M. Gronenborn, G. C. K. Roberts and T. Q. Tran, <u>Br. J. Pharmacol.</u>, 1984, <u>81</u>, 309.
(13) S. J. Hammond, B. Birdsall, J. Feeney, M. S. Searle, G. C. K. Roberts and H. T. A. Cheung, <u>Biochemistry</u>, 1987, <u>26</u>, 8585.

(14) J. R. P. Arnold, S. J. B. Tendler, J. A. Thomas, B. Birdsall, J. Feeney and G. C. K. Roberts, in preparation.
(15) B. Birdsall, J. Andrews, G. Ostler, S. J. B. Tendler, J. Feeney, G. C. K. Roberts, R. W. Davies and H. T. A. Cheung, Biochemistry, 1989, 28, 1353.
(16) M. A. Jimenez, J. R. P. Arnold, J. A. Thomas, G. C. K. Roberts, J. Feeney & B. Birdsall, submitted for publication.
(17) S. J. Hammond, B. Birdsall, M. S. Searle, G. C. K. Roberts and J. Feeney, J. Mol. Biol., 1986, 188, 81.
(18) J. Feeney, B. Birdsall, J. Akiboye, S.J.B. Tendler, J.J. Barbero, G. Ostler, J.R.P. Arnold, G.C.K. Roberts, A. Kuhn and K. Roth, FEBS Letts., 1989, in press
(19) E. I. Hyde, B. Birdsall, G. C. K. Roberts, J. Feeney and A. S. V. Burgen, Biochemistry, 1980, 19, 3738.
(20) E. I. Hyde, B. Birdsall, G. C. K. Roberts, J. Feeney and A. S. V. Burgen, Biochemistry, 1980, 19, 3747.
(21) G. M. Clore, A. M. Gronenborn, B. Birdsall, J. Feeney and G. C. K. Roberts, Biochem. J., 1984, 217, 659.
(22) S. J. B. Tendler, R. J. Griffin, B. Birdsall, M.F.G. Stevens, G.C.K. Roberts and J. Feeney, FEBS Lett., 1988, 240, 201.
(23) H. T. A. Cheung, M. S. Searle, J. Feeney, B. Birdsall, G. C. K. Roberts, I. Kompis and S. J. Hammond, Biochemistry, 1986, 25, 1925.
(24) M. A. Jimenez, J. R. P. Arnold, J. A. Thomas, G. C. K. Roberts, J. Feeney & B. Birdsall, in preparation.
(25) J. A. Thomas, J. Andrews, J. R. P. Arnold, G. C. K. Roberts, B. Birdsall and J. Feeney, unpublished work.
(26) J. Andrews, C.A. Fierke, B. Birdsall, G. Ostler, J. Feeney, G.C.K. Roberts, and S.J. Benkovic, Biochemistry, 1989, in press.
(27) B. Birdsall, A. S. V. Burgen, E. I. Hyde, G. C. K. Roberts and J. Feeney, Biochemistry, 1981, 20, 7186.
(28) L. Cocco, J. P. Groff, C. Temple, jr., J. A. Temple, R. E. London and R. A. Blakley, Biochemistry, 1981, 20, 3972.
(29) G. C. K. Roberts, J. Feeney, A. S. V. Burgen and S. Daluge, FEBS Lett., 1981, 131, 85.
(30) K. Hood and G. C. K. Roberts, Biochem. J., 1978, 171, 357.
(31) A. W. Bevan, G. C. K. Roberts, J. Feeney and L. Kuyper, Eur. J. Biophys., 1985, 11, 211.
(32) J. R. P. Arnold, J. De Graw, L.-Y. Lian and G. C. K. Roberts, unpublished work.

(33) A. M. Gronenborn, B. Birdsall, E. I. Hyde, G. C. K. Roberts, J. Feeney and A. S. V. Burgen, Mol. Pharmacol., 1981, 20, 145.
(34) B. Birdsall, A. W. Bevan, C. Pascual, G. C. K. Roberts, J. Feeney, A. M. Gronenborn and G. M. Clore, Biochemistry, 1984, 23, 4733.
(35) B. Birdsall, A. M. Gronenborn, E. I. Hyde, G. M. Clore, G. C. K. Roberts, J. Feeney and A. S. V. Burgen, Biochemistry, 1982, 21, 5831.
(36) B. Birdsall, J. De Graw, J. Feeney, S.J. Hammond, M.S. Searle, G.C.K. Roberts, W.T. Colwell, and J. Crase, FEBS Lett., 1987, 217, 106.
(37) B. Birdsall, J. Feeney, S. J. B. Tendler, S. J. Hammond and G. C. K. Roberts, Biochemistry, 1989, 28, 2297.
(38) M.S. Searle, M.J. Forster, B. Birdsall, G.C.K. Roberts, J. Feeney, H.T.A. Cheung, I. Kompis and A.J. Geddes, Proc. Natl. Acad. Sci. USA, 1988, 85, 3787.
(39) G. H. Hitchings and B. Roth, in 'Enzyme Inhibitors as Drugs', ed. M. Sandler, Macmillan, London, 1980, p. 263.
(40) P. A. Charlton, D. W. Young, B. Birdsall, J. Feeney and G. C. K. Roberts, J. Chem. Soc. Chem. Comm., 1979, 922.
(41) P. A. Charlton, D. W. Young, B. Birdsall, J. Feeney and G. C. K. Roberts, J. Chem. Soc. Perkin I, 1985 1349.

# Crystallographic Comparison of Penicillin-recognizing Enzymes

By J. R. Knox and J. A. Kelly

DEPARTMENT OF MOLECULAR AND CELL BIOLOGY AND INSTITUTE OF MATERIALS SCIENCE,
THE UNIVERSITY OF CONNECTICUT, STORRS, CT 06269, USA

What stereochemistry distinguishes a penicillin-inhibited enzyme from a penicillin-destroying one? Crystal structures of each type of enzyme are now available, and we can begin to search for features which determine the unique yet closely related chemistry catalysed by each enzyme. The enzyme inhibited by β-lactam antibiotics (penicillins and cephalosporins)

Penicillins

Cephalosporins

is a bacterial cell wall synthesizing enzyme which normally acts on a lysyl-D-alanyl-D-alanine peptide precursor.[1] After breaking the D-ala-D-ala peptide bond, it adds an amino acceptor to form a new peptide bond which crosslinks glycan strands into the cell wall network. The two steps catalysed by this serine-containing DD-peptidase are:

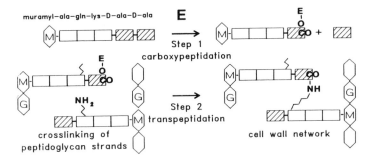

β-Lactams inhibit the first step, presumably because they resemble the peptide substrate. The covalent β-lactamoyl intermediate formed with a reactive serine is rather stable ($t_{1/2}$=hours). The intermediate is ultimately hydrolysed by water, and the active enzyme is regenerated.

On the other hand, β-lactamase[2] (commonly called penicillinase or cephalosporinase, depending on specificity) is able to hydrolyse β-lactam antibiotics at rates approaching 1000 $sec^{-1}$. In this enzyme the acyl intermediate is therefore quite unstable. Can we explain why both enzymes recognize β-lactam antibiotics, but produce very different kinetic results. Another question, possibly addressed with these new crystallographic results, is why the β-lactamase, thought to have evolved from the DD-peptidase, has lost the ability to recognize the lys-D-ala-D-ala cell wall peptide.

We will first describe, individually, the two molecules, then bring them together for a comparative analysis of the three-dimensional geometry of their β-lactam binding sites.

## The Penicillin-Inhibited Enzyme

We have reported the x-ray crystal structure of the D-alanyl-D-alanine cleaving carboxypeptidase-transpeptidase (DD-peptidase)[3,4] from <u>Streptomyces</u> R61 at a resolution of 2.25 A. This 38,000-dalton penicillin-sensitive enzyme consists of eight helical segments and a beta sheet of five antiparallel strands; an alpha-carbon atom stereo-plot is shown in Figure 1a.

The binding site for β-lactams, as determined experimentally from crystallographic maps of binary complexes (Figure 1b), is in the center of Figure 1a at the left edge of the beta sheet. This site is presumably also the site of binding for cell wall peptides prior to their crosslinking, though crystallographic mapping of peptide binding has been difficult because of the catalytic activity of the crystalline enzyme.

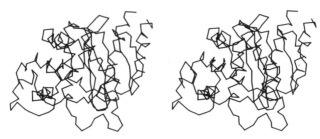

__Figure 1a__    Partial alpha-carbon trace of DD-peptidase of
                S. R61.  N-terminus is on  right-most  helix.

__Figure 1b__    Experimental electron density maps of cephalo-
                thin (left) and (2,3)-α-methylene penicillin G
                (right) as acyl complexes with DD-peptidase.

A view of the binding site region (Figure 1c) shows
polypeptide    segments    which    interact    with    the
antibiotic.   Most important are  a  helix  (residues
62-77), whose N-terminus is near the β-lactam, and a
beta-strand (298-303) which contains several conserved

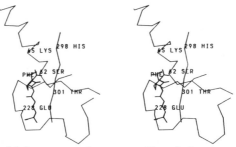

__Figure 1c__   Peptide segments near the β-lactam binding site
                in DD-peptidase.

amino acids and along which lies the β-lactam molecule. Both the inhibitor and, we think, the cell wall peptide substrate employ an antiparallel hydrogen bonding with this beta strand in order to align themselves for nucleophilic attack by serine 62 on the helix. The serine's position at the N-terminus of the helix macrodipole will tend to lower the pk$_a$ of the hydroxyl proton, making it more labile than usual. This proton lability may be assisted by a positively-charged lysine 65, whose amino group lies within 3 A of the hydroxyl group. A more detailed atomic-level picture of these and other enzyme segments will be discussed after we have seen the β-lactamase structure.

## The Penicillin-Destroying Enzyme

The crystal structure of the β-lactamase of Bacillus licheniformis 749/C has been established at 2A resolution. This 29,000-dalton enzyme catalyses the hydrolysis of β-lactams to inactive products:

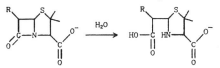

and thereby creates formidable therapeutic problems of longstanding concern. The hydrolysis reaction is thought to proceed via an acyl intermediate which results from the attack of a serine on the carbonyl of the β-lactam bond. Some β-lactamases lack the reactive serine, using instead a mechanism based on a catalytic zinc center. These so-called Class B β-lactamases will not be discussed further, as they appear to have little relation to the penicillin-inhibited DD-peptidase.

An alpha-carbon tracing of the β-lactamase is seen in Figure 2a, which shows an arrangement of helices and beta-sheet very similar to that seen for the DD-peptidase in Figure 1a. The catalytic site is between a helical cluster and the inner edge of the beta-sheet. We will focus only on those components of the enzyme involved in binding the β-lactam substrate. Figure 2b shows that three segments of the enzyme are especially important: the 70-85 helix containing the reactive serine 70, the 233-240 beta-strand, and the 164-175 helix at the bottom of the binding site. Other short peptide segments will be introduced in the Comparison Section below. We found that the beta-strand contains a tripeptide sequence (lysine 234,

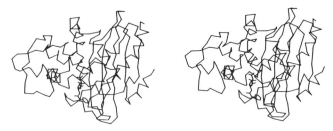

Figure 2a   Alpha-carbon trace of B. lichen. β-lactamase

Figure 2b   Peptide segments near the β-lactam binding
            site in β-lactamase

serine 235, glycine 236) which recurs in equivalent
form in all serine β-lactamases and in all cell wall
synthesizing DD-peptidases.[6] These beta-strand
residues, while probably not catalytically functional,
are possibly important for orientation of the
antibiotic prior to attack by the serine 70. The
helical peptide segment at the bottom of Figure 2b
contains an invariant glutamic acid 166 which may
assist deacylation by directing a water molecule to the
acyl serine bond. Mutagenesis experiments are underway
in many laboratories to test these hypotheses.[7,8]

## Comparison of the Two Enzymes

Similarities in Recognition of β-Lactams. The
relative positions of the helix and beta strand, common
to each structure, are quite similar in the two
molecules (Figure 3a). These and other components of
the antibiotic binding site in the DD-peptidase and
β-lactamase are shown in Figures 3b and 3c. The
catalytic serine and nearby lysine are on the same side
of the helix in a ser-x-x-lys sequence. The
beta-strand in each molecule contains the same or

similar amino acids in equivalent positions: his-thr-gly
in the DD-peptidase and lys-thr-gly in the β-lactamase.
The histidine or lysine group, and the neighboring threonine group, are likely involved in charge-charge or
hydrogen bond interaction with the requisite
carboxylate group on the thiazole ring of β-lactams.
This carboxylate group is thought to be the counterpart
of the C-terminal carboxylate in the lys-D-ala-D-ala
peptide substrates of the DD-peptidase (see Figure 4).

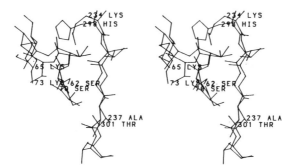

Figure 3a    Superposition of helix and beta strand from
             binding sites of DD-peptidase and β-lactamase.
             Ser 70 and lys 234 belong to β-lactamase.

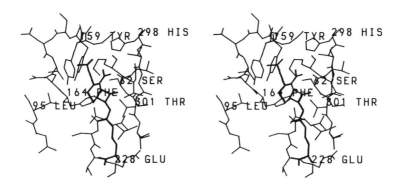

Figure 3b    Environment of cephalosporin C in the
             DD-peptidase of S̲. R61.

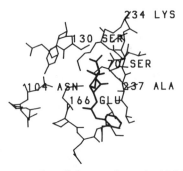

<u>Figure 3c</u>   Environment of benzylpenicillin in the
             β-lactamase of <u>B</u>. <u>licheniformis</u> 749/C.

We think it is the alpha face of the β-lactam
which is presented to the reactive serine, as suspected
also from theoretical calculations and mechanistic
studies.[10]   The serine hydroxyl would be about 2 A from
the carbonyl carbon of the β-lactam ring.   In
preparation for nucleophilic attack on the carbonyl,
either enzyme can polarize the carbonyl bond by forming
hydrogen bonds with two mainchain amides of serine
62(70) and of the beta-strand amino acid at position
301(237), just after the conserved tripeptide sequence
his(lys)-thr-gly.   In this scheme, the side chain of
301(237) need not be, and is not, conserved in either
family of enzymes.   While this kind of
hydrogen-bond-induced oxyanion hole has been seen in
the serine proteases of the chymotrypsin family, we
should note that the penicillin-recognizing enzymes do
not contain a catalytic histidine within a charge
transfer triad asp-his-ser.   A full catalytic mechanism
is still sought for the two enzymes described here,[1,2]
and direct proton transfer to the β-lactam has not been
ruled out.   In spite of this uncertainty, we assume
that the acylation pathway followed by each
penicillin-recognizing enzyme[11,12,13] proceeds via a
tetrahedral intermediate:

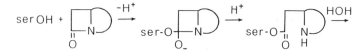

With conventional crystallographic measurements it is
difficult to observe the short-lived complexes,

especially with β-lactamase. The more stable covalent acyl complex with the DD-peptidase has been seen for cephalosporin C,[4,9] (2,3)-α-methylene benzyl penicillin, a benzyl monobactam[14], cephalothin and cefotaxime.[15] Complexation with a cyclobutanone, lacking the β-lactam nitrogen, may permit visualization of the tetrahedral hemiketal complex.

Differences in Catalysis of β-Lactams. To understand why these two enzymes differ so significantly in their ability to deacylate the β-lactamoyl complex is one goal of these crystallographic studies. The DD-peptidase complex is only slowly deacylated, a situation which makes β-lactams excellent inhibitors of cell wall synthesis. The β-lactamase complex is very rapidly deacylated, a situation leading to the efficient destruction of β-lactams in the cell's environment.

Examination of Figures 3b and 3c shows that when the common helix and strand are ignored, certain differences exist in the two binding sites. For example, in the DD-peptidase a phenylalanine (164) lies very near the alpha face of the β-lactam. Even if the orientation of the β-lactam changes slightly upon acylation by the serine 62, as we see experimentally,[4,9] the large, hydrophobic phenylalanine ring will still be near the acyl bond, and it could conceivably hinder approach of the water molecule needed for deacylation. In marked contrast, the β-lactamase (Figure 3c) has at this spatial position the carboxylate group of glutamic acid 166, which is invariant in all β-lactamases and presumably aids entry of the water molecule. Another difference which makes one catalytic site more hydrophilic than the other is seen at the lower left of Figures 3b and 3c, where leucine 95 in the DD-peptidase is replaced by asparagine 104 in the β-lactamase. Also possibly playing a role in β-lactamase catalysis is serine 130 at the top of the binding site. This polar group is common to all Class A β-lactamases, but it may not exist in this position in the DD-peptidases.

Remaining Questions About Substrate Recognition. As we mentioned in the introduction, the β-lactamase has no ability to recognize and react with the L-lys-D-ala-D-ala substrate of its likely ancestor, the DD-peptidase.[16,17] As a new gene product, the nascent β-lactamase presumably lost this peptide specificity as it began to optimize turnover of β-lactams.

It is clear, through our modelling studies, that
the cell wall peptide can be crudely positioned in the
binding site of the β-lactamase so that the C-terminal
carboxylate and the susceptible peptide carbonyl bond
can interact with the β-lactamase in the same general
way they interact with the DD-peptidase. This is
because the conformations of the inhibitor and
substrate, as they were modelled to the enzymes, are
quite similar (Figure 4), though the two carboxylate
groups differ in position by more than 1 A.

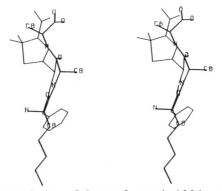

Figure 4    Conformations of benzyl penicillin and
            L-lysyl-D-alanyl-D-alanine as modelled to
            β-lactamase and DD-peptidase, respectively.

In the DD-peptidase, favorable hydrophobic interactions
with leucine 95, tyrosine 159, threonine 162 and
phenylalanine 164 may be required to position the
methyl groups of the peptide substrate. A search of
the β-lactamase binding site shows that it contains
more hydrophilic groups than does the DD-peptidase, but
none of them appear to block the methyl groups. The
substrate's L-lysyl substituent may be another
discriminating factor, for the side chain has more
space and complementary interactions at the bottom of
the DD-peptidase than it has in the β-lactamase. That
this β-lactamase geometry is not sufficiently selective
to prevent entry, hydrolysis or aminolysis of acyclic
depsipeptides has recently been demonstrated by
Pratt.[18] It is likely that the low barrier to rotation
about the ester bond, compared to the higher barrier in
a peptide bond, permits enzyme-induced configurations
in the analog not attainable with the cell wall
peptide.

REFERENCES

1. J.M. Frere and B. Joris, Crit. Rev. Microbiol., 1985, 11, 299.
2. A. Coulson, Biotechnol. Genet. Eng. Rev., 1985, 3, 219.
3. J.A. Kelly, J.R. Knox, P.C. Moews, G.J. Hite, J.B. Bartolone, H. Zhao, B. Joris, J.M. Frere and J.M. Ghuysen, J. Biol. Chem., 1985, 260, 6449.
4. J.R. Knox, J.A. Kelly, P.C. Moews, H. Zhao, J. Moring, J.K.M. Rao, J.C. Boyington, O. Dideberg, P. Charlier and M. Libert, In 'Three Dimensional Structure and Drug Action', (Y. Iitaka and A. Itai, eds.) University of Tokyo Press, Tokyo, 1987, pp. 64-82.
5. P.C. Moews, J.R. Knox, O. Dideberg, P. Charlier and J.M. Frere, submitted for publication, 1989.
6. B. Joris, J.M. Ghuysen, G. Dive, A. Renard, O. Dideberg, P. Charlier, J.M. Frere, J.A. Kelly, J.C. Boyington, P.C. Moews and J.R. Knox, Biochem. J., 1988, 250, 313.
7. S.C. Schultz and J.H. Richards, Proc. Natl. Acad. Sci. USA, 1986, 83, 1588.
8. P.J. Madgwick and S.G. Waley, Biochem. J., 1987, 248, 657.
9. J.A. Kelly, J.C. Boyington, P.C. Moews, J.R. Knox, O. Dideberg, P. Charlier, M. Libert, J.P. Wery, C. Duez, B. Joris, J. Dusart, J.M. Frere and J.M. Ghuysen, In 'Frontiers of Antibiotic Research', (H. Umezawa, ed.) Academic Press, Tokyo, 1987, pp. 327-337.
10. D.B. Boyd, In 'Chemistry and Biology of β-Lactam Antibiotics', Vol. 1 (R.B. Morin and M. Dorman, eds.) Academic Press, New York, 1982, pp. 437-545.
11. A.L. Fink, Pharma. Res., 1985, 55.
12. R.F. Pratt, In 'Design of Inhibitors as Drugs', (M. Sandler and H.J. Smith, eds.) Oxford University Press, 1988.
13. J.R. Knowles, Accts. Chem. Res., 1985, 18, 97.
14. J.A. Kelly, J.R. Knox, H. Zhao, J.M. Frere and J.M. Ghuysen, submitted for publication, 1989.
15. G. Lowe and S. Swain, In 'Recent Adv. Chem. β-Lactam Antibiotics', Royal Society of Chemistry, London, 1985, pp. 209-221.
16. J.A. Kelly, O. Dideberg, P. Charlier, J.P. Wery, M. Libert, P.C. Moews, J.R. Knox, C. Duez, C.I. Fraipont, B. Joris, J. Dusart, J.M. Frere and J.M. Ghuysen, Science, 1986, 231, 1429.
17. R.F. Pratt and C.P. Govardhan, Proc. Natl. Acad. Sci. USA, 1984, 81, 1302.
18. C.P. Govardham and R.F. Pratt, Biochem., 1987, 26, 3385.

# Studies in the β-Lactam Area

By D. W. Young

SCHOOL OF CHEMISTRY AND MOLECULAR SCIENCE, UNIVERSITY OF SUSSEX, FALMER, BRIGHTON BN1 9QJ, UK

β-Lactam antibiotics are the most studied group of antibacterial compounds and range from the classical compounds of the penicillin and cephalosporin series to the more recently discovered thienamycins, nocardicins and monobactams. These compounds are effective by interfering with bacterial cell wall synthesis. They are recognised by and react with key enzymes in this process. These enzymes are known as penicillin binding proteins or PBPs and transpeptidases are one class of PBPs.

Transpeptidases are enzymes which complete the cross-linking process by which a glycine residue attached to one polysaccharide/peptide chain displaces the terminal D-alanine residue of a D-Ala-D-Ala moiety of a second such chain as shown in Figure 1 below.

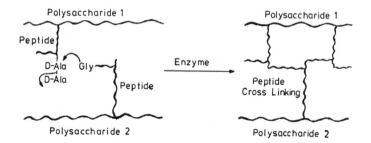

**Figure 1** Transpeptidase catalysed cross-linking in bacterial cell wall synthesis

It has been suggested[1] that recognition of the
D-Ala-D-Ala residue by the enzyme occurs with the
dipeptide residue in conformation (1) so that the enzyme
will also recognise unnatural substrates such as
penicillin (2) which have key functional groups in similar
environments to those of the dipeptide moiety (1).

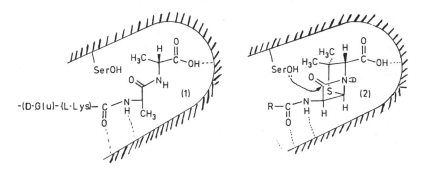

**Figure 2** Binding of natural substrate (1) and penicillin
(2) at the active site of a penicillin binding protein

The ring strain on the amide bond of the β-lactam
structure makes it more reactive than the natural peptide
substrate. This reactivity is enhanced by the difficulty
experienced by the lone pair in the 4.5-bicyclic
penicillin structure (2) in becoming p-hybridised and
therefore taking part in amide resonance. The β-lactam
carbonyl group is therefore very electrophilic and will
react with the side chain hydroxyl group of a serine
residue in the enzyme as shown below. A covalent bond is
formed as in (5) and the enzyme is inactivated and can no
longer catalyse synthesis of bacterial cell wall. It is
easier for the lone pair in the 4.6-bicyclic cephalosporin
structure to participate in amide resonance but the
enamine system shown in (3) competes with this and so
makes the carbonyl group more electrophilic.

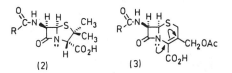

(2)          (3)

Transpeptidase / Carboxypeptidase

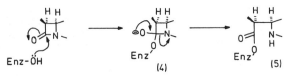

Enz-ÖH        Enz' (4)        Enz (5)

Resistance to penicillins and cephalosporins is exhibited by bacterial strains which have developed enzymes known as β-lactamases which show homology to transpeptidases. These act upon β-lactam antibiotics by a similar mechanism to the PBPs but can hydrolyse the enzyme-bound intermediate (5) thus regenerating enzyme and inactivating the antibiotic. Some β-lactam-containing compounds such as clavulanic acid (6) can form intermediates such as (7) which are not hydrolysed and so they are β-lactamase inhibitors.

β-Lactamase

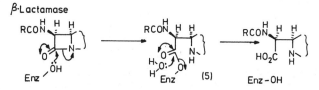

Enz        Enz (5)        Enz-OH

β-Lactamase inhibitor

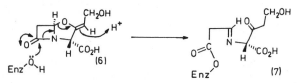

Enz (6)        Enz (7)

The importance of ring strain in β-lactam antibiotics to their biological activity has long been recognised and Morin[2] has drawn a parallel between ring strain as measured by the stretching frequency of the β-lactam carbonyl group in the infrared spectrum of the ester and the antibacterial activity of the free acid for a series of such compounds. At the outset of our work we were interested to examine the effect of additional small rings on the reactivity of such compounds and set out to synthesise compounds with a small ring fused to the "top" of the β-lactam structure as in (8) and (9). Further, although the 4.6- and 4.5-systems of cephalosporins and

penicillins respectively are known, the 4.4-system (**10**)
has yet to be prepared and should be more reactive than
either of the known β-lactam antibiotics.

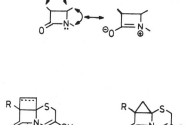

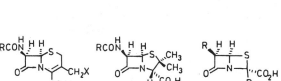

(8)    (9)

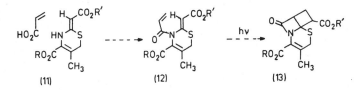

(10)

The 4.4.6 system (**8**) seemed the most approachable of
these targets as we foresaw very straightforward routes to
it.  The thiazines (**11**) were well known[3] and we felt that
formation of a peptide bond to form (**12**) followed by a $2\pi_S$
+ $2\pi_S$ photochemical cyclisation, which we had already used
to prepare fused 4.4-β-lactams[4] might lead directly to the
system (**13**) by route A.  The thiazines (**11**) have enamine
character however and so might react with the acrylates in
a Michael fashion before completion of the peptide bond.
If this were to be the case then modification of the
acrylate might allow preparation of a pyridone (**14**) in one
step as in route B.  Pyridones are known to undergo
photochemical electrocyclic reaction to yield 4.4-systems[5]
and so the analogue (**15**) would be accessible.

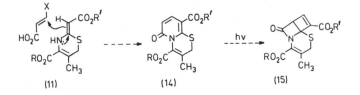

(11)                    (14)                    (15)

It therefore appeared that we might have a
straightforward two-step preparation of our first target
molecule.  In the event, the task proved more difficult
but this gave rise to some extremely interesting chemistry.

Synthesis of the thiazines (11) was readily achieved
by literature methods[3] although in some cases we obtained
the azepinedione (17) as a by-product.  This would appear
to result from Michael reaction of the first-formed
thiazine (11) with the enone (16) followed by completion
of the peptide linkage.

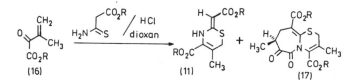

(16)                    (11)                    (17)

The observation of the Michael reaction above would
suggest that the $\pi 2_s + \pi 2_s$ route might not be feasible and
indeed reaction of thiazines (11) with acrylates gave a
series of fused dihydropyridones (18) in excellent yield.
This made the pyridone route B a likely proposition and in
fact reaction of the thiazine with propiolic acid gave the
pyridone (19), albeit in reduced yield.  The low yield
seemed to be caused by formation of the *trans* adduct (20)
as a by-product.  This would result from the initial
Michael step and, unlike the alternative *cis* intermediate,
would not be able to cyclise to the desired pyridone.

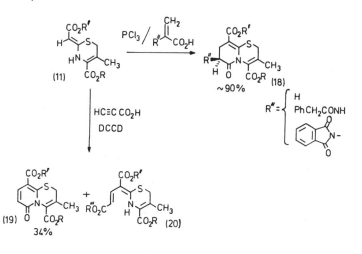

We had now achieved the first step of our two-step synthesis, Michael reaction of the enamine preceding amide formation. When β,β-dimethylacrylate was used in the reaction, a mixture of the two geometrical isomers of the enamine acylation product (**21**) was obtained. Presumably steric hindrance had blocked Michael attack in this case. Use of α-haloacids also resulted in the enamine acylation products (**22**) and (**23**).

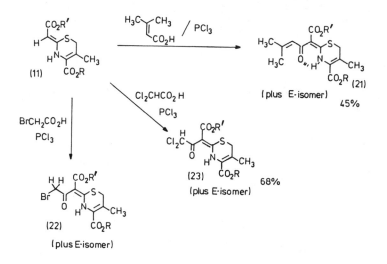

Since it was necessary for the enamine system of the thiazine to have a β-hydrogen if the intermediate imine were to tautomerise to the enamine needed for cyclisation, it was of interest to see what would happen if this position were blocked. The blocked thiazine (**24**) was therefore prepared by our standard synthesis and, when this was reacted with the acrylates (**25**), compounds were obtained which had all of the attributes of the pyridopyrrole system (**26**).

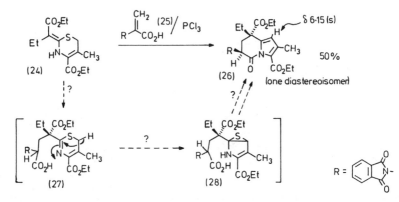

It would appear that, after the initial Michael attack to yield the intermediate imine (**27**), the acidic hydrogen adjacent to sulphur in the thiazine is lost to yield the thiirane (**28**). Extrusion of sulphur and ring closure would then yield the product (**26**).

Having achieved the first step in the two step synthesis and having examined the scope of the method, we were now ready to photolyse the pyridone (**19**) to obtain the target molecule (**15**). In the event, photolysis gave an excellent yield of a new compound which evidently retained the pyridone ring. X-ray crystallography revealed the product to be the pyridothiazetidine (**29**) and photolysis of related compounds such as the dihydropyridone (**18**, R=H) gave good yields of analogous compounds (e.g. (**30**) ).

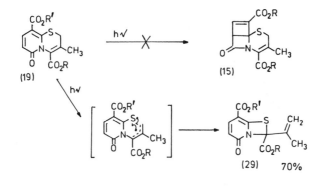

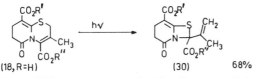

The thiazetidines were relatively novel heterocyclic compounds and some attempts were made to study their chemistry. When the thiazetidine diethyl ester (**29**, R=R'=Et) was hydrogenated using palladium on charcoal or platinum as catalyst an interesting rearrangement occurred to yield the thiazolidine (**31**) as proved by a relatively straightforward independent synthesis using penicillamine (**32**) and the iminoether (**33**) as shown. In an attempt to assess the mechanism of this reaction using deuterium gas, it was seen that deuterium entered both methyl groups to almost the same extent. It is of note that with the photolysis of the thiazine (**19**) to thiazetidine (**29**) followed by the hydrogenolytic rearrangement to (**31**) we had formally converted a compound containing a cephalosporin-type thiazine ring into a compound with a penicillin-type thiazolidine system.

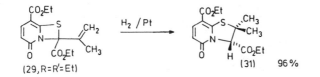

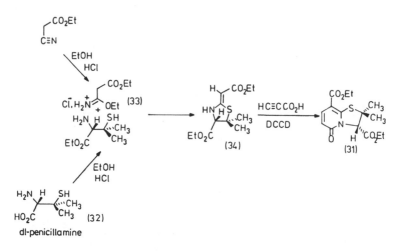

When the methyl ester (35) was used in the catalytic
hydrogenation, the yield of the thiazolidine (36) was
reduced, the anhydropenicillin analogue (37) being
obtained as by-product.

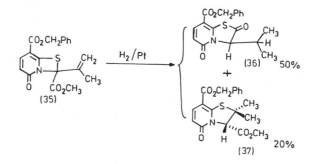

Changing to a homogeneous catalyst caused a formal
[1.3]-rearrangment of the thiazetidine (29) to the
thiazine (14) whilst oxidation with a peracid gave more of
the [1.3]-rearranged sulphone (38) than of the formally
more likely [2.3]-rearranged oxathiazepine (39).

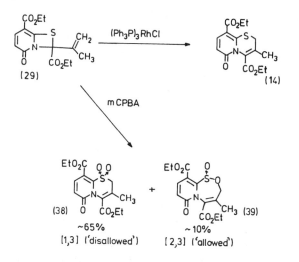

(29)     (14)

(38) ~65% [1,3] ('disallowed')     (39) ~10% [2,3] ('allowed')

Having discovered a synthetic route to thiazetidines from thiazines, we were intrigued by the possibility of directly converting a cephalosporin-type compound (**40**) to the fused 4.4-target compound (**41**). Previous work had suggested[6] that in cephalosporins with a side-chain amide at C-7, the 7-amide would become involved in the photochemical reaction and so the chiral 7-unsubstituted compound (**42**) was prepared. Photolysis in methanol gave the methanol adduct (**44**) which might suggest the intermediacy of the desired 4.4-system (**43**), although the fact that (**44**) was racemic might not be in keeping with this mechanism. No isolable products were obtained using other solvents except when the triplet sensitiser acetone was used in tetrahydrofuran. In this case, stereospecific addition of acetone and tetrahydrofuran to the double bond gave the adduct (**45**). The structures of both photoproducts (**44**) and (**45**) were confirmed by X-ray structure determination.

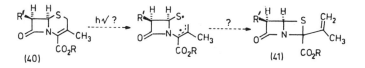

(40)     (41)

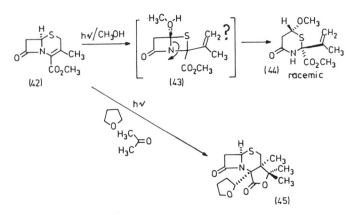

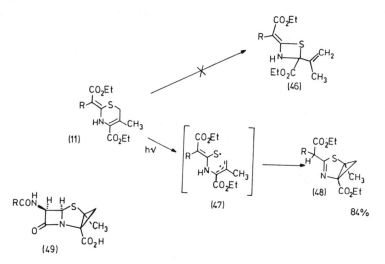

Having failed to prepare the 4.4-system directly, we considered the possibility of first preparing a thiazetidine (e.g. (**46**) ) and subsequently forming the β-lactam ring. However, when the monocyclic thiazine (**11**) was photolysed, an excellent yield of the bicyclic thiazoline (**48**) was obtained. Although this reaction suggests a facile route to known β-lactamase inhibitors such as (**49**),[7] the photolysis has evidently involved an alternative mode of bond formation of intermediate (**47**) or its imine tautomer from that observed with the N-acylthiazines,

An unstable monocyclic thiazetidine (**50**) was now
obtained by reduction of the thiazine (**11**) with
cyanoborohydride. Formylation to the more stable N-acyl
analogue (**51**) and photolysis gave the monocyclic compound
(**52**).

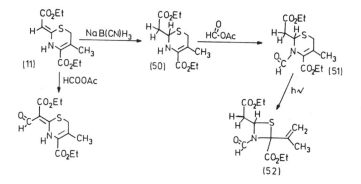

This is the current status of our attempts to prepare
the β-lactam-thiazetidine series and in the remainder of
the lecture I would like to concentrate on our attempts to
prepare the cephalosporin analogue (**8**) bridged across C-6
and C-7 with a four-membered ring. When the
pyridothiazine (**11**) was converted to the sulphoxide (**53**)
using *meta*-chloroperbenzoic acid, photolysis went in a
different way from before yielding the
pyrido-oxathiazoline (**54**) whose structure was verified by
X-ray crystallography.

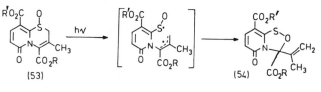

66%

In all of our photolysis reactions the first bond to
break always appeared to be the allylic sulphur-carbon
bond. Further, the pyridine was extremely unreactive and
a search of the literature revealed that very few
successful photolyses of pyridones substituted with
electron withdrawing groups had been achieved.[5] We
therefore decided to convert the ester group on the
pyridone to an electron donating group and to deactivate

the carbon-sulphur bond by reducing the double bond in the
thiazine ring.

We prepared the ethyl benzyl ester (**55**) by our
standard synthesis and preferentially removed the benzyl
ester by reaction with boron tribromide. In our earlier
studies, the acid (**56**) was converted to the alcohol (**57**)
by borohydride reduction of the mixed anhydride. The
$\Delta^3$-double bond was now resistant to catalytic reduction
but rearrangement to the $\Delta^2$-olefin (**58**) gave a compound
which could be hydrogenated to the reduced thiazine (**59**).

A later colleague found that reduction of the mixed
anhydride of acid (**56**) gave the reduced thiazine (**59**)
directly and indeed was never able to prepare the alcohol
(**57**). We can only assume that very different samples of
sodium borohydride were used in these reactions. Both
samples of reduced thiazine (**59**) were identical and were
assigned the *cis* stereochemistry on the basis of [1]H nmr
spectroscopic coupling constants.

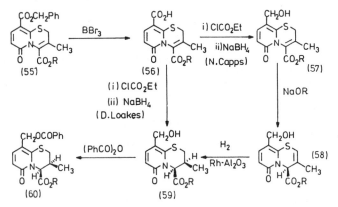

We have prepared both the O-benzoyl and O-methyl
derivatives of the modified pyridone (**60**) and photolysis
of these yielded products with no pyridone absorption in
the [1]H nmr spectra but with olefinic absorption at δ 6.4
ppm. These compounds reverted to the starting pyridones
on standing and so they were hydrogenated *in situ* to yield
the products (**61**) and (**62**) respectively. These had infra
red bands consistent with a reduced cephalosporin system
and indicating little additional ring strain. The COSY
nmr spectrum of the benzoyl derivative (**61**) shown in
Figure 3 was consistent with both sets of coupled protons
in the structure.

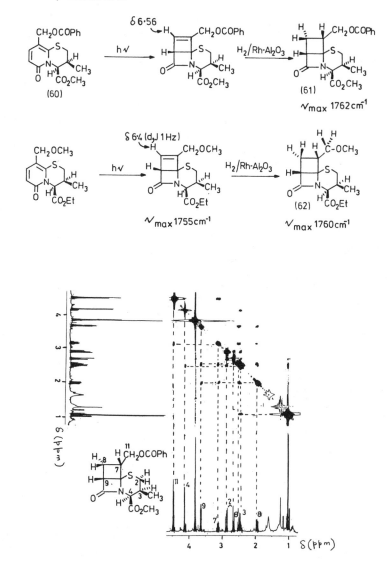

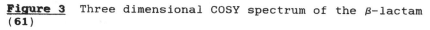

**Figure 3** Three dimensional COSY spectrum of the β-lactam (61)

We have therefore synthesised a modified version of
one of our target molecules.  The infra red spectrum
indicates that it is not as strained as we had hoped
although the $\Delta^3$-olefin may have proved more interesting in
this respect.  So far very few free acids have been
prepared so that it is not surprising that none of the
compounds has biological activity.

Current work in the area involves collaboration with
Dr. Brain Spratt on $\beta$-lactamases modified by site specific
mutagenesis and studies of their interaction with modified
$\beta$-lactam compounds.

The work I have discussed has been achieved through
the efforts of three hardworking and talented young
colleagues, Richard McCabe, Nigel Capps and David Loakes.
It has been supported by SERC and ICI Pharmaceuticals
Division through CASE studentships.  I should like to
acknowledge Dr. Gareth Davies of ICI Pharmaceuticals who
has been involved in the work from the outset and Dr.
Peter Hitchcock who did all of the X-ray structure
determinations.  Some of the work has been published in
the form of preliminary communications.[8-13]

REFERENCES

1.  D.J. Waxman, R.R. Yocum and J.L. Strominger,
    Phil. Trans. R. Soc. Lond. B, 1980, 289, 257.
2.  R.B. Morin, B.G. Jackson, R.A. Mueller, E.R.
    Lavagnino, W.B. Scanlon and S.L. Andrews,
    J. Amer. Chem. Soc., 1969, 91, 1401.
3.  (a) D.M. Green, A.G. Long, P.J. May and
    A.F. Turner, J. Chem. Soc., 1964, 766;
    (b) S.H. Eggers, V.V. Kane and G. Lowe, J. Chem.
    Soc., 1965, 1262.
4.  P.K. Sen, C.J. Veal and D.W. Young, J. Chem.
    Soc., Perkin Trans. I, 1981, 3053.
5.  see for example (a) R.C. De Selms and
    W.R. Schleigh, Tetrahedron Letters, 1972, 3563;
    (b) L.J. Sharp and G.S. Hammond, Mol.
    Photochem., 1970, 2, 225; (c) W.L. Dilling,
    N.B. Tefertiller and A.B. Mitchell, Mol.
    Photochem., 1973, 5, 371; (d) C. Kaneko,
    K. Shiba, H. Fujii and Y. Momose,
    J. Chem. Soc., Chem. Commun., 1980, 1177; and
    references in these papers.
6.  Y. Maki and M. Sako, J. Amer. Chem. Soc., 1977,
    99, 5091.

7.  D.D. Keith, J. Tengi, P. Rossman, L.Todaro and
    M. Weigele, Tetrahedron, 1983, 39, 2445.
8.  R.W. McCabe, D.W. Young and G.M. Davies, J. Chem.
    Soc.,Chem. Commun., 1981, 395.
9.  P.B. Hitchcock, R.W. McCabe, D.W. Young and
    G.M. Davies, J. Chem. Soc., Chem. Commun., 1981,
    608.
10. N.K. Capps, G.M. Davies, P.B. Hitchcock,
    R.W. McCabe, and D.W. Young, J. Chem. Soc.,
    Chem. Commun., 1982, 1418.
11. N.K. Capps, G.M. Davies, P.B. Hitchcock,
    R.W. McCabe, and D.W. Young, J. Chem. Soc.,
    Chem. Commun., 1983, 199.
12. N.K. Capps, G.M. Davies and D.W. Young,
    Tetrahedron Letters, 1984, 4157.
13. N.K. Capps, G.M. Davies, P.B. Hitchcock, and
    D.W. Young, J. Chem. Soc., Chem. Commun., 1985,
    843.

# The Interaction of Antiviral Agents with Human Rhinovirus-14 Capsid Protein

By Guy D. Diana, Adi M. Treasurywala, Thomas R. Bailey, and R. Christopher Oglesby

STERLING RESEARCH GROUP, RENSSELAER, NY 12144, USA

## 1 INTRODUCTION

Over the past several years, a series of compounds encompassed in structure I have been shown to demonstrate potent activity against a wide spectrum of rhino- and

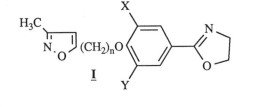

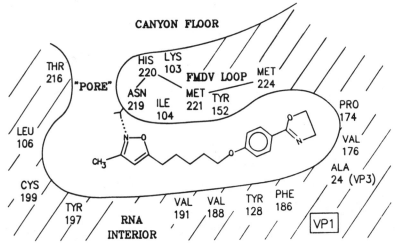

Figure 1 A schematic representation of the HRV-14 compound-binding site.

enteroviruses.[1-5] These compounds inhibit viral replication by preventing uncoating in the case of rhinovirus-2 and polio 2,[6,7] and by blocking adsorption of the virus to the cell in the case of rhinovirus-14.[8] Although the mode of action has been clearly established, the mechanism by which the drugs bind to the viral capsid was unknown.

Recently, the 3-dimensional structure of HRV-14 was elucidated[9] and subsequently, several oxazolinylphenyl isoxazoles I, bound to the capsid protein of HRV-14 were examined by X-ray crystallography.[10,11] The results of these studies have allowed use to examine interactions of these compounds with the virus on a molecular level and also to develop some hypotheses concerning the mode of binding of these compounds to the viral capsid.

The compound-binding site resides on the surface of the viral capsid protein VP 1 below a depression, referred to as a canyon, and which is considered to be the putative cell receptor binding site. The site consists of a hydrophobic pocket ~25Å long and 8 to 10Å wide, which undergoes a conformational change of between 3-5Å following insertion of the compounds (Figure 1). It has been shown that this conformational change extends to the canyon or cell receptor binding site in HRV-14 which may account for the inhibition of adsorption of the virus to the cell by these compounds[8]. With the X-ray coordinates in hand, a systematic evaluation of the molecular interaction of portions of the molecule with the receptor was performed and the effect of these interactions on activity examined.

## 2 ENANTIOMERIC EFFECTS

One of the initial compounds examined by X-ray crystallography was compound II which inserts into the pocket of VP 1 with the isoxazole end residing in the interior of a β-barrel. The methyloxazolinyl moiety is located below the canyon floor with the nitrogen of the oxazoline ring within hydrogen bonding distance of Asn 219. Compound II has an asymmetric center at the 4-position of the oxazoline ring; it was subsequently determined that the S-isomer was 10 times more active than the R. This result was consistent with the fact that preferential binding of the S-isomer occurred at the binding site in VP 1.[11]

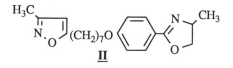

$\underline{\mathbf{II}}$

Additional homologs (Table 1) were prepared and tested against HRV-14 in order to determine the extent of this enantiomeric effect, and to examine the interaction of the extended chain with residues within the pocket. In every case, the S-isomer was more active than the R. Optimum activity was achieved with the ethyl and propyl homologs.

Table 1   Comparative Evaluation of Enantiomers Against HRV-14

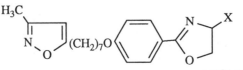

MIC(μMol)

| X | S | R |
|---|---|---|
| CH₃ | 0.056 | 0.56 |
| C₂H₅ | 0.03 | 0.16 |
| n-C₃H₇ | 0.03 | 0.18 |
| i-C₃H₇ | 0.08 | 1.57 |
| n-C₄H₉ | 0.15 | 1.31 |

\*Absolute Configuration

MOLECULAR GRAPHICS

The interaction of the R and S conformers in the binding site were analyzed by performing an energy profiling study using the X-ray structure of II in the HRV-14 binding site as a starting point, and examining the interaction of the compounds in Table 1 with residues within 8Å. The purpose of this study was to determine the location of energy minima as the oxazolinyl ring was rotated through 360°. All calculations were performed on a VAX 11/785. Hydrogen atoms were removed and charges were set on the atoms in the pocket and on the compounds. The intermolecular van der Waals energy was calculated using a 6-12 function and rotations about the bond between the phenyl and oxazoline rings were performed in increments of 10°. The results are shown in Figure 2. The S-methyl and ethyl homologs showed valley points at 10-30° while the corresponding R-isomers exhibited a broad valley in this area which was not low lying, very similar to the desmethyl compound which was less active. The gem dimethyl compound produced a pattern similar to the S-methyl homolog which is consistent with the comparable level of activity of these two compounds.

The coordinates of the S-conformer of II were displayed on an Evans and Sutherland graphics device using the program Chem-X. The difference in activity of the S and R conformers could be rationalized by the observed hydrophobic

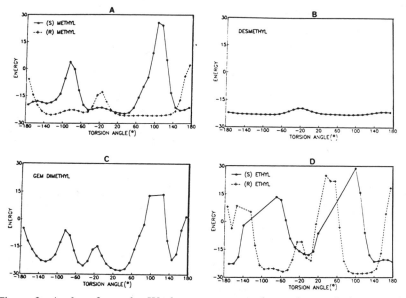

<u>Figure 2</u> A plot of van der Waals energy versus the torsion angle between the oxazoline and phenyl rings of compound II.

interaction of the S-methyl group with a pocket formed by Leu 106 and Ser 107, (Figure 3). The R isomer on the other hand is pointing away from this pocket. Assuming the same torsion angle between the phenyl and oxazoline rings for the R isomer as was determined for the S isomer by X-ray crystallography, the former would interact unfavorably with the carbonyl group of Asn 198, (Figure 4). The results of this study suggested that a rigidly confined stable conformation with a twist angle of 10-30°, is conducive to high levels of activity. This was consistent with the results of the X-ray data for compound II which revealed a torsion angle of 10-15° between these rings.

In performing the energy profiling studies, hydrophobic interactions were taken into account and considered as the major contributor to the binding energy.

## 3 STRUCTURE-ACTIVITY STUDIES

X-ray crystallography studies have been performed on several compounds in the isoxazole series I with both 5 and 7 carbon chains connecting both ends of the molecule.[11] A series of homologs in the disoxaril series and analogs containing a chlorine on the phenyl ring (Table 2) were evaluated against HRV-14 in an effort to establish a structure-activity relationship between chain length and activity. In both series, a chain length of 7 carbon atoms was required for optimum activity. The significance of these findings will become more obvious further on in the discussion.

Table 2    In Vitro Activity Against HRV-14

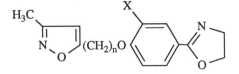

MIC (μMol)

| X = | n = | 4 | 5 | 6 | 7 | 8 |
|---|---|---|---|---|---|---|
| | H | NA | 0.73 | 2.9 | 0.41 | 3.92 |
| | Cl | 9.2 | 2.41 | 3.86 | 1.06 | 14.32 |

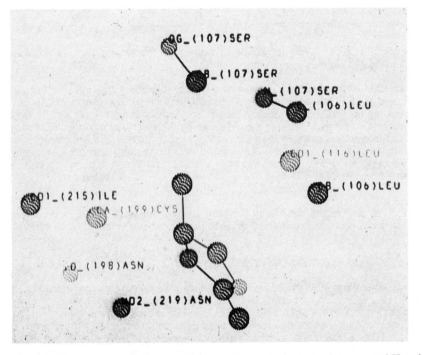

Figure 3  The interatomic distances between the methyl group of compound II and $C_B$ of Ser[107] (4.1Å), $C_B$ of Leu[106] (4.7Å), and $C_A$ of Cys[199] (4.95Å)

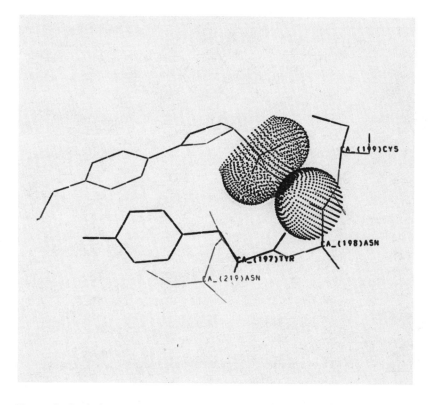

Figure 4   Steric interaction of the methyl group of the R isomer of compound II with the carbonyl of Asn[198].

## QSAR Analysis

We have also examined the effect of substituents on the phenyl ring on activity against HRV-14. The thirteen compounds shown in Table 3 were screened, and the antiviral results subjected to a regression analysis to establish a correlation between activity and some physicochemical parameters. Log p (an indicator of lipophilicity), molecular weight, (an indicator of bulk), and $sigma_m$, (representing an inductive effect), were used. The data was analyzed using a step-wise regression analysis. The results in Table 4 indicate a poor correlation between log[1/MIC], where MIC is the minimum inhibitory concentration, and log p (R=.58), and no correlation with MW (R=.29), or $sigma_m$ (R=.14). However, a combination of log p and MW resulted in a correlation of R=.80 (see equation 5). The results suggest that molecular weight (bulk) may have a substantial negative effect on activity.

Table 3   In Vitro Activity Against HRV-14

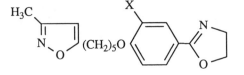

| X | MIC($\mu Mol$) | X | MIC($\mu Mol$) |
|---|---|---|---|
| H | 0.73 | CHO | 2.34 |
| $CH_3$ | 0.61 | Cl | 2.40 |
| $C_2H_5$ | 1.14 | $CF_3$ | 3.04 |
| $CH_3O$ | 1.66 | $NH_2$ | 3.34 |
| F | 1.68 | $CH_3CO$ | 7.30 |
| $NO_2$ | 1.78 | $CH_2OH$ | 15.7 |
| Br | 2.04 | | |

Table 4   Regression Analysis

| 1. | log[1/MIC] | vs | logp | n=13 | R=.58 | S=.32 |
|---|---|---|---|---|---|---|
| 2. | " | vs | $\sigma_m$ | | R=.14 | S=.41 |
| 3. | " | vs | MW | | R=.29 | S=.39 |
| 4. | " | vs | logp + $\sigma_m$ | | R=.41 | S=.35 |
| 5. | log[1/MIC] = .476 logp - 4.09 MW+ 1.49 | | | | R=.80 | S=.26 |

## 4 A MODEL FOR HRV-14 ACTIVITY

The elucidation of the 3-dimensional structure of HRV-14, the identification of the compound-binding site on the capsid protein, and the X-ray structures of several compounds bound to HRV-14 has allowed for the development of a model representing the requirements for activity for this series of compounds. The model was developed by examining 7 active compounds (Table 5) and 7 inactive compounds (Table 6). In the former series, the conformation of compounds 1, 3, 4 and 5 bound to HRV-14 was determined by X-ray crystallography. Compound 2 and 6 were constructed using the program SYBYL from the X-ray conformation of 1; and 7 was constructed from compound 4. The assumption was made, based on the X-ray crystallographic results on a variety of compounds in this series[11], that compounds 1, 2 and 6 would assume the same orientation while 5 and 7 would have the opposite orientation when bound to HRV-14. Since the determination of the coordinates for the inactive compounds by X-ray methods was not possible, compound 4 (Table 5) was used as a template for all of the inactive compounds in Table 6. The composite of active and inactive structures generated was then compared using volume maps comprised of van der Waals surfaces. The volume maps were examined and differences between the active and inactive structures noted. The results of this comparison revealed two major differences:

1. Inactive compounds display excess bulk around the phenyl ring. Although some bulk in this area appears desirable, exceeding certain limitations results in inactive compounds, possibly due to spatial constraints within the compound binding site.

2. The active compounds occupy space in the binding site beyond that of the inactive compounds. The most active compounds (1 and 3) extend well into the pore area of the binding site suggesting a space filling requirement in this area. Compound 1 and 3 are in the opposite orientation of each other such that the methyl group on the oxazoline ring of compound 1 extends into the pore while the hydroxylated side chain of 3 is in the same area (Figure 1).

The conclusions drawn from this model substantiate the results of the SAR studies on the chain length previously described[13] where a chain length of 7 carbon atoms appears to satisfy the space-filling requirements of the model. The negative effect of molecular weight on activity resulting from the QSAR emphasizes the bulk limitations. The poor correlation of overall lipophilicity with activity also substantiates this observation since it is not merely the presence or absence of hydrophobic bulk in the molecule but the actual location of that bulk which appears to correlate with activity.

The importance of lipophilicity and bulk on activity was supported by the use of a QSAR method called CoMFA (Comparitive Molecular Field Analysis), (Tripos Associates). The concept embodied in this program is that differences in an objective property are often related to differences in the shapes of the non-covalent field surrounding molecules in question. Steric and electrostatic fields are taken into consideration and are sampled periodically within a region and the magnitude of these fields are put into a QSAR table. The QSAR is evaluated by its cross-validated $r^2$ value and if acceptable, the CoMFA QSAR is visualized by 3-D contouring.

Structures obtained from X-ray analysis of eight compounds bound to HRV-14 and whose activity against HRV-14 was determined were used for this analysis. The structures were extracted into a SYBYL database and charges were calculated for them using the AM1 Hamiltonian in a single point calculation. The CoMFA analysis consists of placing these molecules collectively into an imaginary

Table 5    Compounds Active Against HRV-14

MIC(μMol)

Cmpd. #                                                          HRV-14

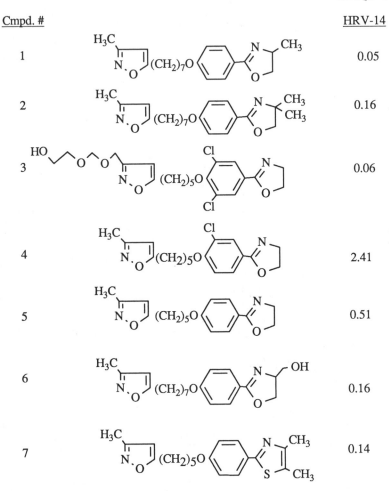

| Cmpd. # | HRV-14 |
|---|---|
| 1 | 0.05 |
| 2 | 0.16 |
| 3 | 0.06 |
| 4 | 2.41 |
| 5 | 0.51 |
| 6 | 0.16 |
| 7 | 0.14 |

box and then the box into grid points which are evenly spaced. The molecules are then placed individually into the box and values of a given property (or set of properties) contributed by the molecule are evaluated at each grid point. From this analysis a table is generated consisting of a dependent variable (MIC) and values at each grid point. The data in the table is suited for a classical QSAR analysis and the program seeks a statistically significant correlation between some property value somewhere in space around a set of molecules and a dependent variable, in

<u>Table 6</u>  Compounds Inactive Against HRV-14

Cmpd. #

| | |
|---|---|
| 8 | |
| 9 | |
| 10 | |
| 11 | |
| 12 | |
| 13 | |
| 14 | |

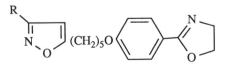

Table 7  In Vitro Activity Against HRV-14

| R | MIC($\mu$Mol) |
|---|---|
| $CH_3$ | 1.27 |
| $C_2H_5$ | 0.43 |
| $C_3H_7$ | 0.16 |

this case, biological activity. A Partial Least Squares (PLS) analysis is performed on all entries in the table with cross validation. The significant conclusions drawn from this analysis were the following:

1. There was no statistically significant correlation between the attraction for a unit positive or negative charge around the molecules and biological activity.

2. There was a statistically significant correlation between the presence of bulk in some regions (particularly in the pore region) and the absence of bulk in other regions, and the antiviral activity of the compounds. These results are illustrated graphically by a plot of the so-called standard deviation residuals in a contour map around the molecules.These results are consistent with the data obtained from direct inspection of the virus pocket and from other computational experiments.

In order to further substantiate the conclusions drawn from the volume map study as well as the results from the CoMFA analysis concerning the importance of the location of bulk on activity, three compounds shown in Table 7 were synthesized and tested against HRV-14. The prediction would be that by the addition of bulk at the terminus of the isoxazole ring the antiviral activity should accordingly increase. The results in Table 7 confirm these expectations where there is an increase in activity as the length of the hydrocarbon chain is increased from methyl to propyl. It is interesting to note that the propyl homolog has the same molecular weight as the seven carbon homolog shown in Table 2 but with enhanced activity.

The results of these studies suggest that the use of the three dimensional structure of rhinoviruses may be a useful tool for the design of more potent antirhinovirus agents. As the structure of other serotypes becomes available, comparable studies will be performed.

REFERENCES

1.  G.D. Diana, M.A. McKinlay, M.J. Otto, V. Akullian, R.C. Oglesby, J. Med. Chem. 1985 28, 1906 (1985).
2.  M.J. Otto, M.P. Fox, M.J. Fancher, M.F. Kuhrt, G.D. Diana, and M.A. McKinlay, Antimicrob. Ag. Chemother. 1985 27, 883.
3.  C.M. Wilfert, J.R. Zeller, L.E. Schauber, R.L. McKinney, Abstract No. 430, 24th Interscience Conference on Antimicrobial Agents and Chemotherapy, Washington, D.C. 1984.
4.  G.D. Diana, R.C. Oglesby, V. Akullian, P.M. Carabateas, D. Cutcliffe, J.P. Mallamo, M.J.Otto, M.A. MicKinlay, E.G. Maliski, and S.J. Michalec, J. Med. Chem., 1987 30 383.
5.  G.D. Diana, D. Cutcliffe, R.C. Oblesby, M.J. Otto, J.P. Mallamo, V. Akullian, and M.A. McKinlay, J. Med. Chem., 1989 32, 450.
6.  M.P. Fox, M.J. Otto, W.J. Shave, and M.A. McKinlay, Antomicrob. Ag. Chemother., 1986 30, 110.
7.  C.M. Wilfert, J.R. Zeller, L.E. Schauber and R.L. McKinney, Abstract No. 430, 24th Interscience Conference on Antimicrobial Agents and Chemotherapy, Washington, D.C., 1984.
8.  D.C. Peavaer, M.J. Fancher, P.J. Felock, M.A. Rossmann and F.J. Dutko, J. Virol., 1989, in press.
9.  M.J. Rossmann, E. Arnold, J.W. Erickson, E.A. Frankenberger, J.P. Griffith, H.J. Hecht, J.E. Johnson, G. Kramer, M. Luo, A.G. Mosser, R.R. Reuckert, B. Sherry and G. Vriend, Nature, 1985 317, 145.
10. T.J. Smith, M.J. Kremer, M. Luo, G.Vriend, E. Arnold, G. Kamer, M.G. Rossmann, M.A. McKinlay, G.D. Diana, and M.J. Otto, Science, 1986, 233, 128.
11. J. Badger, I. Minor, M.J. Kremer, M.A. McKinlay, G.D. Diana, F.J. Dutko, M. Fancher, R.R. Reuckert, and B.A. Heinz, B.A., Proc. Natl. Acad., Sci., 1988, 85, 3304.
12. G.D. Diana, M.J. Otto, A.M. Treasurywala, M.A. McKinlay, R.C. Oglesby, E.G. Maliski, M.G. Rossmann, and T.J. Smith, J. Med. Chem., 1988, 31, 540.
13. G.D. Diana, M.A. Mckinlay, C.J. Brisson, E.S. Zalay, J.V. Miralles and U.J. Salavador, J. Med. Chem. 1985, 28, 749.

# Protein Crystallography, Computer Graphics, and Sleeping Sickness

By W. G. J. Hol, R. K. Wierenga*, H. Groendijk, R. J. Read,
A. M. W. H. Thunnissen, M. E. M. Noble†, K. H. Kalk,
F. M. D. Vellieux, F. R. Opperdoes*, and P. A. M. Michels*

BIOSON RESEARCH INSTITUTE, UNIVERSITY OF GRONINGEN, NIJENBOGH 16, 9747 AG GRONINGEN,
THE NETHERLANDS
†EUROPEAN MOLECULAR BIOLOGY LABORATORY, POSTFACH 10.2209, 6900 HEIDELBERG, FRG
*RESEARCH UNIT FOR TROPICAL DISEASES, ICP, AVENUE HIPPOCRATE 74, B-1200 BRUSSEL,
BELGIUM

## 1 INTRODUCTION: TOWARDS MORE RATIONAL DRUG DESIGN

The development of new pharmaceutically active
compounds is a major challenge to science, involving a wide
range of disciplines. In the past, new drugs have been
discovered or developed in quite different ways. Sometimes,
naturally occurring compounds already in use for centuries
proved to be useful in modern times. An example is,
qinghaosu or artemisinin. This anti-malarial compound,
isolated from a herb, was already described in Chinese
pharmacopedia more than a millennium ago although only
recently the molecular structure of the active compound was
elucidated [1]. In other cases, new drugs were obtained
after screening a large number of different compounds in an
assay system. More rational are approaches which start from
deep insight into crucial metabolic pathways of pathogenic
organisms. This can involve a study of the substrates and
products of these processes followed by synthesis of
substrate analogues, transition state analogues and suicide
inhibitors. This approach has proven to be of immense value
[see e.g. 2].

Today we are witnessing an explosion of insight into
biological processes, down to the atomic level. This is due,

on the one hand, to gene cloning and expressing techniques,
which provide large quantities of protein material and, on
the other hand, to protein crystallography which is the main
technique in determining three-dimensional structures of
proteins. This is adding a new dimension to rational drug
design: it becomes possible to start from the three-
dimensional structure of well-chosen target proteins and use
this knowledge to design selective inhibitors which prevent
proper functioning of these proteins. This is by no means
straightforward. Hence, a rational drug design cycle [3] is
envisaged which consists of building up a database of
structures of the target protein complexed with a variety of
compounds (Fig. 1). In this manner information is gained

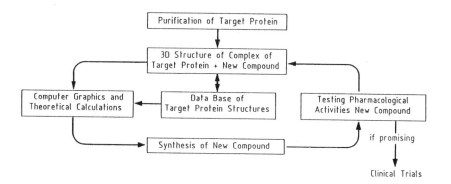

<u>Figure 1</u>   The protein-structure based drug design cycle

about binding modes of small molecules and flexibility of
the large protein molecule. We have embarked, in an
expanding collaboration with other groups, upon a project
aiming at protein-structure-based design of new sleeping
sickness drugs.


## 2 SLEEPING SICKNESS, OR AFRICAN TRYPANOSOMIASIS

Sleeping sickness is considered by the World Health
Organisation as one of the major current tropical infectious
diseases and is included in the UNDP/World Bank/WHO Special
Program for Research and Training in Tropical Diseases.
African trypanosomiasis is caused by a unicellular
flagellate, *Trypanosoma brucei*, which is transmitted by the
tse-tse fly. The parasite has a number of very unusual
biochemical features such as:

- the possession of a uniform coat of variable surface glycoproteins [4] which makes it very difficult to develop vaccines for prevention of the disease;
- the occurrence of trypanothione reductase instead of glutathione reductase [5];
- editing of mitochondrial RNA, a proces that somehow results in the production of RNA molecules which differ in nucleotide sequence in coding regions from the DNA templates [6];
- an extremely rapid glycolysis a major part of which is occurring inside special microbody-like organelles, called the glycosomes [7].

It is known that interference with glycolysis is a rapid way of killing the parasite *in vivo* [8] as well as *in vitro* [9]. Therefore we are focussing on glycolytic enzymes of *Trypanosoma brucei* to provide a starting point for compounds which interfere with glycolysis in the parasite and not, or to a much lesser extent, in the human host.

## 3 GLYCOSOMAL ENZYMES AND DRUG DESIGN

In the bloodstream form of *T. brucei* the glycosome contains nine enzymes involved in glycolysis and glycerol metabolism. These enzymes can be simultaneously purified by chromatographic techniques [10] and have been subjected to a number of biophysical and enzymological investigations [10-12]. Some characteristics of the proteins involved are shown in Table 1.

A common and important property of these enzymes is that they are all synthesized outside the glycosome and rapidly imported into the organelle after synthesis [13].

A consequence of the data obtained so far is that there are three general possible areas of interference with proper functioning of the above mentioned nine enzymes in the sleeping sickness parasite:
a.    prevent entry into the glycosome;
b.    prevent assembly into multimers (table 1). This assembly is essential for activity - only PGK functions as a monomer;
c.    inhibit catalytic activity by blocking the active site. These areas are by no means strictly mutually exclusive, but for sake of simplicity are treated as such in what follows.

The first of these possible areas of interference, import, suffers from the fact that very little is known about the topogenic signals required for import into

Table 1     Some    characteristics    of   *T. brucei*   glycosomal
            enzymes

| Name | Abbr. | Subunit Mr | Subunits per oligomer | pI | Gene sequence known |
|------|-------|-----------|----------------------|------|--------------------|
| Hexokinase | HK | 49.500 | 6 | 10.0 | - |
| Phosphogluco isomerase | PGI | 52.500 | 2 | 7.5 | [14] |
| Phospho fructo kinase | PFK | 49.000 | 4 | 8.7 | - |
| Aldolase | ALDO | 39.000 | 4 | 9.1 | [15,16] |
| Triose P Isomerase | TIM | 27.000 | 2 | 9.8 | [17] |
| Glyceraldehyde P Dehydr. | GAPDH | 37.000 | 4 | 9.3 | [18] |
| Phospho glycerate kinase | PGK | 48.000 | 1 | 9.4 | [19] |
| Glycerol P Kinase | GK | 41.000 | 2 | 9.0 | - |
| Glycerol P Dehydrogenase | GDH | 33.000 | 2 | 10.0 | - |

glycosomes. When gene sequences of glycosomal enzymes became
available and iso-electric points were established it
appeared that glycosomal enzymes were, with only one
exception, unusually positively charged. Structural analysis
suggested that patches of positive charges might be the
import signal [20]. However, genes of glycosomal and
cytosolic enzymes in *Crithidia fasciculata*, a species
related to *T. brucei*, provided strong evidence that in PGK
the import signal resides in a C-terminal extension [21]. No
other glycosomal enzyme with known sequence does, however,
have a C-terminal extension, but very recently gPGI from *T.
brucei* has been sequenced by Marchand et al. (in
preparation). This enzyme appeared to have an N-terminal
extension with some sequence homology with the C-terminal
extension of gPGK [Hol et al., in preparation]. It might
therefore be that a general glycosomal import signal has
been found, although the positive charges could be
specifically involved in import of enzymes into the
glycosomes of *Trypanosoma brucei*.

The second possible area of interference is prevention
of assembly. Within the framework of our collaborative
project the attention has been concentrated on mimicking β-
turns of gTIM being involved in intersubunit contacts. Cyclo
hexapeptides mimicking the β-turn made by residues 74-77
have been synthesized by Dr. R. Osowski of the University
of Frankfurt after modelling by Dr. K. Müller, Hoffmann-La
Roche in Basle. A second β-turn being mimicked by a cyclo
hexapeptide comprises residues 44-47. Modelling studies have
been carried out to find out what is the best way to link
these two β-turn mimicking peptides. These linked
cyclopeptides are clearly tremendous synthetic challenges
and the results of these approaches have still to be
awaited. It is obvious, however, that knowledge of the

three-dimensional structure of a target protein provides
deep insight into ways by which one can interfere in
principle with the proper functioning of the protein.

The third possible area of interference is selective
inhibition of the catalytic mechanism. It is this issue we
would like to focus on in the remainder of this
contribution.

## 4 CRYSTALLOGRAPHIC STUDIES OF GLYCOSOMAL GLYCERALDEHYDE-3-PHOSPHATE DEHYDROGENASE (gGAPDH) FROM *T. BRUCEI*

Read et al. [22] were able to obtain two crystal forms
of gGAPDH. For data collection synchrotron radiation was
essential in order to obtain a resolution better than 3 Å
for crystal form II. Using this data, Read et al.
(unpublished results) were able to solve the structure of
this trypanosomal enzyme using the molecular replacement
method [23] and the structure of *B. stearothermophilus* GAPDH
[24] as a starting model. It appeared that 6 subunits, or 1½
tetramers, of gGAPDH were present in the asymmetric unit.
This amounts to 240,000 daltons. While refinement of this
crystal form is progressing, we are exploring possibilities
for inhibitor binding studies to gGAPDH. In collaboration
with Dr. J. Hadju, University of Oxford, Laue films of
gGAPDH in crystal form I have been recorded. So far, data
out to 3.2 Å have been obtained but data processing is still
under way. We hope to be able to obtain higher resolution
data in the future.

With the amino acid sequence obtained by Michels et al.
[18] it is already possible to make general statements
regarding the type of molecules which can become selective
inhibitors of the glycosomal enzyme. Comparison of the NAD
binding environment in GAPDH, and using the human and *T.
brucei* sequences [25,18] it appeared that most residues in

Table 2    Residues near the adenine moiety of NAD differing
           between human and glycosomal GAPDH.

| T. brucei | Human |
|-----------|--------|
| Asn 7 | Asp 7 |
| Val 36 | Asn 33 |
| Asn 39 | Phe 36 |
| Gln 90 | Glu 78 |
| Asn 92 | Asp 80 |
| Leu 112 | Val 100 |

the neighbourhood of NAD are identical, except near the adenine binding pocket. Here substantial sequence differences occur, as listed in table 2. Hence, NAD analogues with variations in the adenine moiety are potential candidates for selective interference of glycolysis in *T. brucei* via inhibition of gGAPDH.

## 5 CRYSTALLOGRAPHIC STUDIES OF GLYCOSOMAL TRIOSE PHOSPHATE ISOMERASE (gTIM) FROM *T. BRUCEI*

gTIM crystals could be grown in the presence of 2.4 M ammonium sulphate [26]. The asymmetric unit contains one dimer. Only 600 μg of protein was required to solve the structure of this protein at 2.4 Å resolution [27]. Subsequently a 1.9 Å resolution dataset has been collected using synchrotron radiation. The structure is nearing the end of the refinement process, with a current R-factor of 19.3% for 37568 reflections between 6 Å and 1.83 Å resolution. In this model 286 water molecules have been located, whereas in the active site of subunit-2 a sulphate

Table 3    Structure    determination    of    gTIM    with    three
            inhibitors

| | Compounds | | |
|---|---|---|---|
| | 3-phosphono-propionic acid (3PP) | glycerol-3 phosphate (G3P) | 3-phospho glycerate (3PGA) |
| $K_i$ | 27 mM | 0.6 mM | 1.3 mM |
| Inhibitor Concentration in mother liquor | 75 mM | 6 mM | 7 mM |
| Number of unique reflections obtained | 10,879 | 18,064 | 18,169 |
| Maximum resolution | 2.8 Å | 2.3 Å | 2.3 Å |
| $R_{diff} = \dfrac{\Sigma|F_{complex}-F_{nat}|}{\Sigma|F_{nat}|}$ | 0.194 | 0.182 | 0.166 |
| R-factor after refinement | 28% | 25% | 23% |

ion is bound. Apparently due to crystal packing effects the accessibility of the two active sites is different. Because of this the conformation of a flexible loop, near the active site, is also different in the two subunits. In subunit-1 (no sulphate bound) this loop (residues 170-180) is in an "open" conformation, whereas in subunit-2 (sulphate present) the loop (residues 470-480) is rather flexible.

Crystallographic inhibitor binding studies initially failed because either crystals cracked or did not bind inhibitors at all. Development of a procedure for transferring crystals from 2.4 M ammonium sulphate to solutions of polyethylene glycol [28] solved the latter problem. Using a FAST TV area detector diffractometer, X-ray data for three inhibitors complexed with gTIM are now available (Table 3). After crystallographic refinement and comparison with the native structure, it turns out that a major conformational change took place near the active site. Specifically, the structure of the residues of the flexible loop is now well defined. Several atoms move more than 6 Å with respect to the "open" conformation, while this movement is quite similar for all three inhibitors studied (Table 4).

A quite detailed picture of the binding mode of these inhibitors has already been obtained, as shown for example for 3-phosphoglycerate in Figure 2. The recognition of the inhibitor by the enzyme involves a total of seven hydrogen bonds, one of which is provided by the flexible loop. It is noteworthy that the phosphor atoms of the three inhibitors fall within 0.3 Å from each other, while the sulphur atoms of the sulphate ion as bound in the native structure is within 0.7 Å from this phosphor site. Preliminary soaking

Table 4    R.m.s. backbone shifts (in Å) flexible loop of the second subunit of gTIM upon binding of three inhibitors

|         | 3PP | G3P | 3PGA |
|---------|-----|-----|------|
| Pro-468 | 0.4 | 0.4 | 0.4  |
| Val-469 | 0.5 | 0.6 | 0.8  |
| Trp-470 | 1.4 | 1.3 | 1.6  |
| Ala-471 | 2.5 | 2.1 | 2.5  |
| Ile-472 | 2.3 | 2.0 | 2.3  |
| Gly-473 | 4.2 | 4.2 | 4.6  |
| Thr-474 | 4.2 | 4.1 | 4.5  |
| Gly-475 | 6.7 | 6.1 | 6.5  |
| Lys-476 | 2.3 | 2.1 | 2.2  |
| Val-477 | 2.5 | 1.7 | 1.7  |
| Ala-478 | 0.4 | 0.5 | 0.6  |

experiments indicate that also the binding of a phosphate
ion in the active site favours the "closed" conformation of
the flexible loop. Apparently subtle differences between
the ions bound in the active site have major consequences
for the position and flexibility of the flexible loop.

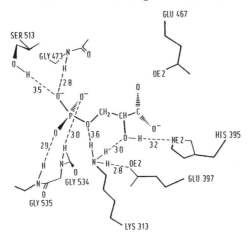

gTIM-3PGA

Figure 2   The  environment  of  3-phosphoglycerate  bound  to
          gTIM


     Initial modelling studies on the basis of the unrefined
open "sulphate" gTIM structure in collaboration with Dr. K.
Müller  had  suggested  that  compounds  like  glycerol-3-
phosphate  ethanolamine,  glycerol-3-phosphate  serine  and
glycerol-3-phosphate  serine-asparagine  might  be  potential
inhibitors of gTIM. For glycerol-3-phosphate serine this has
proved  not  to  be  the  case.  At  that  time  no  accurate
knowledge was available of any inhibitor bound to gTIM, so
that the motion of the flexible loop could not be taken into
account. With our current knowledge we are able to proceed
along two significantly different paths:
(i)  using the "open" structure of gTIM and design compounds
     which fit well into the active site of this open form.
(ii) using  the  "closed"  structure  of  gTIM  inhibitor
     complexes  and  extend  the  size  of  the  inhibitors
     currently known.

     In doing this, it should be kept in mind that near the
active site the sequences of human and trypanosomal TIM are

very much the same. Only at about 10 Å from the "active site phosphate" a tripeptide occurs at the surface of gTIM which deviates from human TIM. These are residues 100-102 being Ala-Tyr-Tyr in the parasite enzyme and His-Val-Phe in the human enzyme. The inhibitors have to be designed in such a way that this difference in sequence and structure is "sensed" by new generations of TIM-inhibitors so that the *T. brucei* enzyme is inhibited several orders of magnitude more effectively than the human enzyme.

This is what we are currently attempting to achieve.

## ACKNOWLEDGEMENTS

We are grateful for stimulating discussions with Bart Swinkels, Piet Borst, Rudolph Osowski, Klaus Müller, Jos van Beeumen and Christophe Verlinde. This work is supported by the WHO Special Program for Tropical Diseases and by a grant of the European Economic Community and the Alberta Medical Heritage Foundation.

## REFERENCES

1.  Cooperative Research Group on the Structure of Qinghaosu, Kexue Tongbau, 1977, 22, 142.
2.  G.B. Elion, Science, 1989, 244, 41-47.
3.  W.G.J. Hol, Angew. Chemie (Engl. Ed.), 1986, 25, 767-778.
4.  K. Vickerman, Nature, 1978, 273, 613-617.
5.  A.M. Fairlamb, P. Blackburn, P. Ulrich, B.T. Chait and A. Cerami, Science, 1985, 227, 1485-1487.
6.  N. Maizels and A. Weiner, Nature, 1988, 334, 469-470.
7.  F.R. Opperdoes and P. Borst, FEBS Lett., 1977, 80, 360-364.
8.  A.B. Clarkson and F.H. Brohn, Science, 1976, 194, 204-206.
9.  A.H. Fairlamb, F.R. Opperdoes and P. Borst, Nature, 1977, 265, 270-27 .
10. O. Misset, O.J.M. Bos and F.R. Opperdoes, Eur. J. Biochem., 1986, 157, 441-453.
11. A.M. Lambeir, F.R. Opperdoes and R.K. Wierenga, Eur. J. Biochem., 1987, 168, 69-74.
12. O. Misset, J. van Beeumen, A.M. Lambeir, R. van der Meer and F.R. Opperdoes, Eur. J. Biochem., 1987, 162, 501-507.
13. D.T. Hart, P. Baudhuin, F.R. Opperdoes and C. de Duve, EMBO J., 1987, 6, 1403-1411.
14. M. Marchand et al., in preparation.

15. C. Clayton, EMBO J., 1985, 4, 2997-3003.
16. M. Marchand, A. Poliszczak, W.C. Gibson, R.K. Wierenga, F.R. Opperdoes and P.A.M. Michels, Mol. Biochem. Parasitol., 1988, 29, 65-76.
17. B.W. Swinkels, W.C. Gibson, K.A. Osinga, R. Kramer, G.H. Veeneman, J.H. van Boom and P. Borst, EMBO J., 1986, 5, 1291-1298.
18. P.A.M. Michels, A. Poliszczak, K.A. Osinga, O. Misset, J. van Beeumen, R.K. Wierenga, P. Borst and F.R. Opperdoes, EMBO J., 1986, 5, 1049-1056.
19. K.A. Osinga, B.W. Swinkels, W.C. Gibson, P. Borst, G.H. Veeneman, J.H. van Boom, P.A.M. Michels and F.R. Opperdoes, EMBO J., 1985, 4, 3811-3817.
20. R.K. Wierenga, B.W. Swinkels, P.A.M. Michels, K.A. Osinga, O. Misset, J. van Beeumen, W.C. Gibson, J.P.M. Postma, P. Borst, F.R. Opperdoes and W.G.J. Hol, EMBO J., 1987, 6, 215-221.
21. B.W. Swinkels, R. Evers and P. Borst, EMBO J., 1988, 7, 1159-1165.
22. R.J. Read, R.K. Wierenga, H. Groendijk, A. Lambeir, F.R. Opperdoes and W.G.J. Hol, J. Mol. Biol., 1987, 194, 573-575.
23. M.G. Rossmann, 1972, Editor of "The Molecular Replacement Method", Gordon Breach, New York.
24. T. Skarzynski, P.C.E. Moody and A.J. Wonacott, J. Mol. Biol., 1987, 193, 171-187.
25. J.Y. Tso, X.-H. Sun, T.-H. Kao, K.S. Reece and R. Wu, Nucleic Acids Research, 1985, 13, 2485-2502.
26. R.K. Wierenga, W.G.J. Hol, O. Misset and F.R. Opperdoes, J. Mol. Biol., 1984, 178, 487-490.
27. R.K. Wierenga, K.H. Kalk and W.G.J. Hol, J. Mol. Biol., 1987, 198, 109-121.
28. H.A. Schreuder, H. Groendijk, J.M. van der Laan and R.K. Wierenga, J. Appl. Cryst., 1988, 21, 426-429.

# A Conformational Approach to the Discovery of Potent Selective Neurokinin Receptor Antagonists

J. R. Brown*, S. P. Clegg†, G. B. Ewan†, R. M. Hagan*, S. J. Ireland*, C. C. Jordan*, B. Porter†, B. C. Ross†, and P. Ward†

†DEPARTMENT OF MEDICINAL CHEMISTRY, GLAXO GROUP RESEARCH LTD., GREENFORD ROAD, GREENFORD, MIDDLESEX UB6 0HE, UK
*DEPARTMENT OF NEUROPHARMACOLOGY, GLAXO GROUP RESEARCH LTD., PARK ROAD, WARE, HERTFORDSHIRE SG12 0DP, UK

## INTRODUCTION

The large and ever increasing number of known peptide hormones and neurotransmitters with diverse physiological actions offers unprecedented opportunities for novel therapeutic strategies and drug discovery. However the pharmacological characterization of peptide receptors and an understanding of the possible pathophysiological roles of neuropeptides requires the development of potent and selective receptor antagonists. The design of such compounds represents a challenging problem in molecular recognition for which there is no established rational approach.

Current progress in the sequencing and structural modelling of G-protein-coupled receptors offers much promise for the future design of receptor antagonists. [This class includes the receptors for the peptides substance P (SP), neurokinin A (NKA), angiotensin II (AII) and vasopressin, amongst others]. Meanwhile, the discovery of peptide antagonists has relied on empirical modifications of the agonist structure, a tedious approach requiring the synthesis of large numbers of analogues. The essential problem is to dissect from structure-activity relationships (SAR) the molecular properties of the ligand (backbone and side-chain conformation, surface topology, electrostatic dipole moment) responsible for (1) receptor recognition or binding, from those resulting in (2) receptor activation leading to signal transduction and

a cellular response.  However, SAR interpretation is
likely to be confounded by the high conformational
mobility of peptides, unless constrained analogues of well-
defined 3-D structure can be studied.

No general rules for the transformation of peptide
agonists into antagonist structures can be deduced from a
survey of the literature.  Indeed, a wide variety of
modifications have successfully given antagonists,
including (1) simple side-chain alterations (eg. [Leu[8]]-
angiotensin II), (2) D-amino acid substitutions (LH-RH,
bradykinin, neurotensin, SP and bombesin antagonists),[1]
(3) backbone modifications [eg. reduced peptide bond replacements (-CH$_2$NH-) in the gastrin antagonist, Boc-Trp-
Met-Ψ(CH$_2$NH)-Asp-Phe-NH$_2$,[2] and in bombesin antagonists[3]].
However, despite the structural diversity of peptide
antagonists, the following general themes have emerged:

(1)  Antagonists interact with receptors in binding modes
 which differ from those of agonists and use  different
structural and conformational features (as revealed by SAR
differences).

(2)  A wider range of structural modifications is
compatible with the strong receptor binding of antagonists
than with the high potency and efficacy of agonists.

(3)  Modifications which restrict the conformational
mobility of a peptide ligand can elicit a change from
agonist to antagonist-type receptor interactions (eg.
[Pen[1]]-oxytocin[4] and the bicyclic vasopressin antagonist,[5]
Figure 1).

(a)  H-Pen-Tyr-Ile-Gln-Asn-Cys-Pro-Leu-Gly-NH$_2$

(b)  Mpa-Phe-Phe-Gln-Asp-Cys-Pro-Lys Gly-NH$_2$

Figure 1     Conformationally constrained oxytocin antagonist
(a) and vasopressin antagonist (b).  (Pen,
-NHCH(CMe$_2$S-)CO-; Mpa, -S-CH$_2$CH$_2$CO-).

The above concepts led us to a speculative but systematic approach to the discovery of antagonists using the localized conformational restriction afforded by ring-fusion to the peptide ligand backbone as a possible means of differentiating the structural requirements for receptor binding and activation.

The present study concerns antagonists of substance P (SP), the first of the so-called 'gut and brain' peptides to be characterized as a biologically active principle, and subsequently identified from its 11-amino acid sequence as a member of the tachykinin family.[6] This homologous group of neuropeptides is characterized by a highly conserved C-terminal region (Phe-X-Gly Leu Met $NH_2$), and is widely distributed throughout the animal kingdom. (Figure 2).

Three mammalian tachykinins [SP, neurokinin A (NKA) and neurokinin B (NKB)] have so far been discovered and there is considerable interest in their possible physiological roles. In particular there is much evidence to suggest that SP and NKA act as mediators of sensory transmission in the spinal cord.[7] Thus, both peptides are localized in unmyelinated slow-conducting primary afferent C-fibres, classically associated with nociception, and are released into the dorsal horn of the spinal cord in res-

Substance P    Arg-Pro-Lys-Pro-Gln-Gln-Phe-Phe-Gly-Leu-Met.NH2

Physalaemin    pGlu-Ala-Asp-Pro-Asn-Lys-Phe-Tyr-Gly-Leu-Met.NH2

Eledoisin      pGlu-Pro-Ser-Lys-Asp-Ala-Phe-Ile-Gly-Leu-Met.NH2

Kassinin       Asp-Val-Pro-Lys-Ser-Asp-Gln-Phe-Val-Gly-Leu-Met.NH2

Neurokinin A       His-Lys-Thr-Asp-Ser-Phe-Val-Gly-Leu-Met.NH2

Neurokinin B       Asp-Met-His-Asp-Phe-Phe-Val-Gly-Leu-Met.NH2

Figure 2   Primary structures of tachykinins.   The conserved residues are underlined. pGlu = pyroglutamic acid.

ponse to noxious peripheral stimuli. Antagonism of the post-synaptic actions of SP and/or NKA at spinal cord receptors might therefore offer a novel approach to analgesia. Tachykinins released from the peripheral terminals of sensory afferent nerves are also strongly implicated in the neurogenic component of various inflammatory disorders. Hence appropriate selective antagonists may be useful for the treatment of wide-ranging disease states, including asthma and ulcerative colitis.

Tachykinin receptors have been classified into three sub-groups (NK-1, NK-2, NK-3) largely on the basis of the activities of selective agonists. Whilst the mammalian tachykinins are relatively non-selective, their rank orders of potency in a range of tissues suggest that SP, NKA and NKB may be the endogenous peptide ligands for the NK-1, NK-2 and NK-3 receptors respectively. Competitive antagonists of SP have been developed, principally by the groups of Folkers[8] and Regoli,[9] through the empirical substitutions of hydrophobic D-amino acids at critical positions (7, 9, 10) in the SP sequence. Since 1981, well over 100 such analogues have been reported, but no real progress in potency or selectivity has been achieved by this approach. For example,[D-Pro[4], D-Trp[7,9,10]]-SP(4-11) does not distinguish between NK-1 and NK-2 receptors, and shows similar affinities in the micromolar range for SP, bombesin and cholecystokinin receptors on guinea pig pancreatic acini.[10] Recently a series of cyclic peptide antagonists, principally showing NK-2 selectivity, has been described[11], but clearly, there is a need for more potent and selective neurokinin antagonists.

Although the detailed molecular mechanisms of receptor mediated signal transduction are essentially unknown, it is generally accepted that agonist binding must induce a conformational change in the extracellular domain which is then transmitted through the transmembrane region of the receptor. When bound to the activated receptor, even a highly flexible peptide agonist might adopt a specific "bioactive" conformation. Other conformers, perhaps energetically disfavoured, could in principle bind to the receptor without initiating a response. Such an intrinsic potential for competitive antagonist activity encoded in the primary sequence of the peptide could be realized in analogues containing conformational constraints designed to prevent access to the agonist bioactive conformation. Conformationally constrained antagonists are also more likely to offer the advantage of receptor selectivity.

RESULTS AND DISCUSSION

In adopting this approach to the discovery of neurokinin
antagonists, the first step was to identify key conformational requirements for the receptor-selective agonist
activity of analogues of SP. The C-terminal "active core"
hexapeptide analogue [Ava[6]]-SP(6-11) was chosen as the
parent compound for conformation-activity studies since it
possesses full agonist activity and is synthetically more
convenient than SP itself. Our attention focussed on Gly[9]
since this residue serves a unique conformational role in
peptides and proteins, for example as a frequent constituent of β-turns. The absence of a side-chain and its
associated steric interactions accounts for the high flexibility of glycine about the backbone torsion angles $\phi$, $\psi$,
as represented by the extensive low energy areas in all
quarters of a conformational energy map for N-acetyl-N'-
methyl-glycinamide (Figure 3). The Gly[9] residue in
[Ava[6]]-SP(6-11) can be regarded as a conformationally
mobile "hinge" linking the hydrophobic Phe[7], Phe[8] and
Leu[10], Met[11] residues which are known to play an important
role in receptor recognition/binding. Moreover, Cascieri
et al. have shown that different carboxy-terminal conformations are recognized by the binding sites for [125]I-
labelled Bolton - Hunter conjugates of SP and eledoisin
respectively.[12] Our overall objective therefore was to
correlate selective agonist and antagonist activities at
NK-1 and NK-2 receptors with the backbone conformation at
position 9 in conformationally restricted analogues of
[Ava[6]]-SP(6-11).

Conformational Requirements for Selective Agonist Activity
at Neurokinin Receptors.

        Replacement of Gly[9] in [Ava[6]]-SP(6-11) by L-residues
(Ala, Pro) increased NK-1 activity (5.8 and 1.9 fold
respectively) and selectivity with respect to the NK-2
receptor (Table 1). In contrast, the corresponding D-
residue substitutions enhanced NK-2 activity and selectivity, D-Pro having a particularly pronounced effect (18
fold increase in potency and 1000 fold increase in
selectivity). Both L-Pro and D-Pro modifications have been
identified by other groups as important structural deter-
minants of the functional activity and receptor binding
affinity of neurokinin agonists[15],[16]. We interpret these
structure-activity relationships to indicate a "partial
bioactive conformation" for NK-1 agonists in which residue
9 occupies the (-, +) or (-, -) regions of the $\phi$, $\psi$ map
(Figure 3). This conclusion follows from the theoretical
and observed preference of L-Ala for these regions of $\phi$, $\psi$

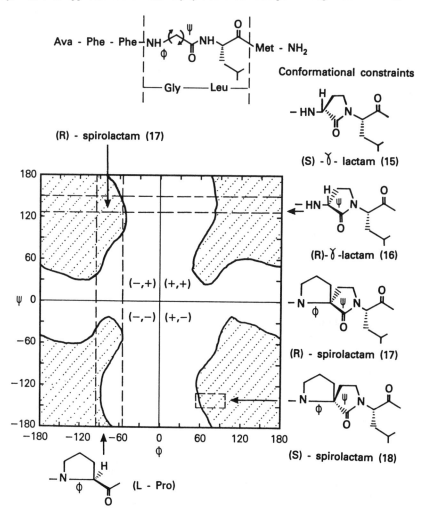


Figure 3   Conformational constraints for Gly[9] in [Ava[6]]-
SP(6-11). Approximate torsional limits allowed
by the constraints (dotted lines) are plotted on
a conformational ($\phi$, $\psi$) energy map for N-acetyl-
N'-methyl-glycinamide (2 Kcal.mol$^{-1}$ contour; low
energy areas shaded).

Table 1    $EC_{50}$ values in NK-1 and NK-2 preparations for
analogues of [Ava[6]]-SP(6-11).

6    7    8    9    10    11
Ava-Phe-Phe-Gly-Leu-Met.$NH_2$

| Comp. No. | Substitution | | Calc. $EC_{50}$ values (nM) | |
|-----------|--------------|---|--------|--------|
|  |  | | NK-1 | NK-2 |
| (1) | $Gly^9$ | | 34 | 1,190 |
| (2) | Ala9 | | 5.9 | 1,860 |
| (3) | D-Ala9 | | 638 | 452 |
| (4) | $Pro^9$ | | 17.7 | 31,500 |
| (5) | $D-Pro^9$ | | 1,960 | 67 |
| (6) | *(R)-γ-lactam$^{9,10}$ | (16) | 75 | 67 |
| (7) | (S)-γ-lactam$^{9,10}$ | (15) | 49,000 | 9,990 |
| (8) | *(R)-spirolactam$^{9,10}$ | (17) | 307 | 1,390 |
| (9) | *(S)-spirolactam$^{9,10}$ | (18) | >98,000 | >80,000 |
| (10) | *(R)-fused lactam$^{9,10}$ | (19) | 478 | 15 |
| (11) | $Trp^7$ | | 529 | 12,500 |
| (12) | $D-Trp^7$ | | >10,000 | 7,530 |
| (13) | $Trp^{11}$ | | 4,380 | >16,000 |
| (14) | $D-Met^{11}$ | | >17,000 | >50,000 |
|  | Substance P | | 4.91 | 167 |
|  | Neurokinin A | | 13.7 | 1.7 |

Agonist activity at NK-1 and NK-2 receptors was
determined from contractile responses of guinea-pig longi-
tudinal smooth muscle (GPI) and rat colon muscularis
mucosae (RC) respectively in the presence of atropine
(1μM), mepyramine (1μM), methysergide (1 μM) and indo-
methacin (1 μM) as described previously.[13]

Experiments were conducted to obtain activities
relative to SP (GPI) or NKA (RC) and $EC_{50}$ values were
calculated by reference to mean values for the standard
agonists.

All compounds except (8),(9) and (10) were
synthesized using the Fmoc - polyamide strategy.[14]

Abbreviation:   Ava, δ-aminovaleryl.

*See Figure 1 for structures of γ-lactam, spirolactam and
fused lactam constraints.

space,[17] and from the absolute exclusion from other regions afforded by the ring constraint in L-Pro. Similarly, the results for D-Ala[9] and D-Pro[9] analogues indicate a Gly[9] bioactive conformation in the (+, +) or (+, -) regions for NK-2 agonists. The proline ring constraints place additional limits (estimated from a survey of crystallographic data[18]) on the $\phi_9$ torsion angles in each bioactive conformation, as depicted in Figures 3 and 4.

Following a similar approach, the torsion angle $\psi_9$ was mapped by incorporating γ-lactam backbone constraints[19] (15) and (16) having (S)- and (R)-configurations respectively. The (S)-γ-lactam-containing peptide (7) was only weakly active at NK-2 receptors and was essentially devoid of agonist or antagonist activity at NK-1 receptors. In contrast, the (R)-γ-lactam analogue (6) was relatively active at both receptor sub-types. The high NK-1 activity ($EC_{50}$ 75 nM) essentially agrees with the functional activity of the corresponding [pGlu[6]]-SP(6-11) analogue reported by Cascieri et al.[12] We speculated that the small (approximately 2-fold) reduction in NK-1 activity compared with the parent structure is probably consistent with a Gly[9] conformation in the (+, +) or (-, +) regions, but perhaps not lying quite within the more limited $\psi$ range allowed by the R-γ-lactam ($\psi$ = +140° ± 10°, estimated from crystallographic data[18]. Conversely, the 18-fold increase in NK-2 activity presumably results from a more accurate mimicry of the NK-2 bioactive conformation.


## Design and Synthesis of Bicyclic Conformational Constraints

The approximate ranges of torsion angles $\phi$ and $\psi$ observed in L-proline and (R)-γ-lactam-containing crystallographic structures respectively are represented in Figure 3 as sets of parallel dotted lines. These intersect in the (-, +) region to enclose an area of essential $\phi$, $\psi$ space that is common to both constraints and therefore contains or approximates to the Gly[9] bioactive conformation at NK-1 receptors. This analysis suggested the design of the (R)-spirobicyclic lactam (17) which combines the structural features of L-Pro and the (R)-γ-lactam and thus constrains both torsion angles within the putative NK-1 bioactive range ($\phi_9$ = -75° ± 20°, $\psi_9$ = + 140° ± 10°).

The conformation-activity data for NK-2 agonists was similarly analysed to predict the fused bicyclic lactam (19) as a conformational constraint for NK-2 agonist activity (Figure 4). This structure effectively combines D-Pro and the (R)-γ-lactam to restrict the position 9 backbone approximately within the bioactive range ($\phi_9$ = +75 ± $20^0$,$\psi_9$ = +140 ± $10^0$).

The (R)-spirolactam was initially synthesized by a non-chiral route which provided the spirolactam pseudo-dipeptide ester analogues (22a,b) <u>via</u> the separable diastereoisomeric intermediates (20a,b) [Scheme 1(a)].

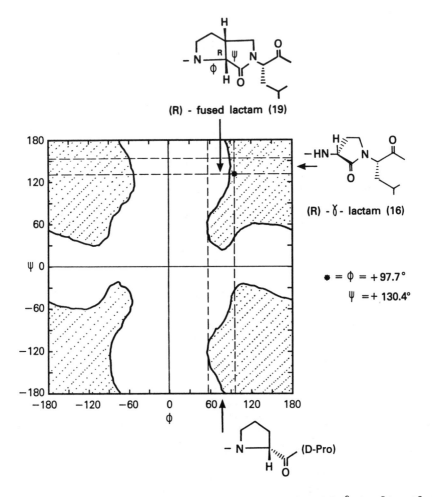

**Figure 4**  Conformational constraints for Gly[9] in [Ava6]-SP (6-11) which are compatible with high NK-2 agonist activity.

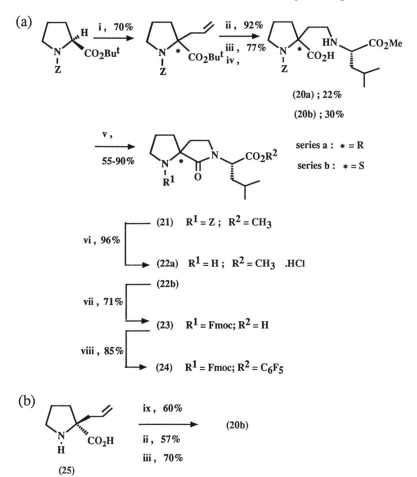

(a)

i, 70% ⟶ ii, 92% ⟶ iii, 77% iv,

(20a) ; 22%

(20b) ; 30%

v, 55-90% ⟶

series a : ∗ = R

series b : ∗ = S

(21) $R^1 = Z$ ; $R^2 = CH_3$

vi, 96%

(22a) $R^1 = H$ ; $R^2 = CH_3$ .HCl

(22b)

vii, 71%

(23) $R^1 = Fmoc$; $R^2 = H$

viii, 85%

(24) $R^1 = Fmoc$; $R^2 = C_6F_5$

(b)

ix, 60% ⟶ (20b)

ii, 57%

iii, 70%

(25)

Scheme 1  Reagents: i, $LiNPr^i_2$, $CH_2=CHCH_2I$, THF; ii, $O_3$, $CH_2Cl_2$ then $Ph_3P$; iii, H-Leu-OMe, THF, mol.sieve 3A then $NaCNBH_3$, MeOH; iv, TFA; v, DIMAPEC, $CH_2Cl_2$; vi, $H_2$, 10% Pd-C, MeOH, HCl; vii, 1N NaOH aq-MeOH, then Fmoc-O-Su, $Na_2CO_3$, $Me_2CO-H_2O$; viii $C_6F_5OH$, N,N'-dicyclohexylcarbodiimide (DCC), dioxan; ix, Z-O-Su, DMF, $Na_2CO_3$; Abbreviations: DIMAPEC, N-dimethylaminopropyl-N'-ethylcarbodimide; Fmoc, 9-fluorenylmethoxycarbonyl; Su, N-succinimidyl; Z, benzyloxycarbonyl.

The configurational assignment was based on the subsequent stereospecific synthesis of the (S)-spirolactam (22b) from D-proline <u>via</u> the chiral α-allyl derivative (25)[20] [Scheme 1(b)]. Both isomers were separately incorporated into [Ava[6]]-SP(6-11) following a standard solution phase peptide synthesis strategy principally using N[α]-Boc-protected amino acid pentafluorophenyl esters.[21] (All peptide analogues reported here were characterized by reverse phase HPLC, amino acid analysis, and FAB mass spectrometry and were 95%+ pure). The synthesis of a related (R)-spirolactam constraint by a different route from that described here has recently been reported.[22]

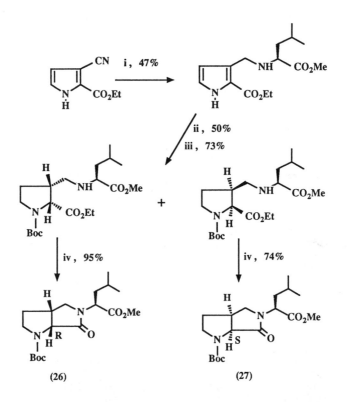

(26)                                        (27)

Scheme 2     Reagents: i, Raney Ni, H-Leu-OMe.HCl, EtOH-H$_2$O; ii, 4-dimethylaminopyridine, Boc$_2$O; iii, H$_2$, 5% Rh-Al$_2$O$_3$, EtOH, 60 psi; iv, NEt$_3$, reflux. Boc = t-butoxycarbonyl.

A non-chiral synthesis afforded both (R)- and (S)-fused lactams as the protected pseudo-dipeptide analogues (26) and (27) [Scheme 2; assignment of absolute configuration from X-ray crystallographic analysis of (26)].

As anticipated from design considerations, the (R)-spirolactam-containing analogue (8) possessed full intrinsic agonist activity in guinea-pig ileum (GPI; NK-1 preparation), albeit about 9-fold less active than the parent hexapeptide. This reduced activity may reflect some departure from the true bioactive conformation as proposed for the monocyclic (R)-γ-lactam analogue (6), but with more serious consequences in the case of the more rigid bicyclic constraint. The (R)-fused bicyclic lactam (19), however, was highly effective as a constraint for the NK-2 bioactive conformation. For example the peptide analogue (10) ($EC_{50}$ 15 nM) approached NKA in NK-2 agonist activity and was approximately 4-fold more selective with respect to NK-1 receptors. Values of the backbone torsion angles calculated from X-ray crystallographic data for intermediate (26) ($\phi$ = +97.7⁰, $\psi$ = +130.4⁰) are represented as a point (*) on the $\phi$, $\psi$ map (Figure 4) and this lies essentially on the boundary of the overlap region for the D-Pro and (R)-γ-lactam constraints. Thus the activities of peptides containing the (R)-spiro- and (R)-fused- bicyclic constraints confirm our previous analysis and unequivocally locate the residue 9 bioactive conformation for NK-1 and NK-2 agonists in the (−, +) and (+,+) $\phi$, $\psi$ regions respectively.

## The development of potent and selective NK-1 antagonists.

Whereas the (R)-spirolactam constraint was compatible with NK-1 agonist activity, the enantiomeric (S)-spirolactam locks the residue 9 backbone in the (+, −) region (Figure 3) and should exclude both NK-1 and NK-2 bioactive conformations, thereby abolishing agonist activity. The (S)-spirolactam analogue (9) was indeed devoid of agonist activity at concentrations up to 30 μM. Moreover, this compound retained sufficient affinity for the NK-1 receptor to act as an antagonist, causing a parallel rightward shift of the SP methyl ester concentration - response curve in the GPI (p$K_B$ 5.6; Table 2) but, even at 100 μM, it failed to antagonize the effects of NKA in the rat colon (NK-2).

In developing the novel lead afforded by compound (9), structural modifications of Phe[7], Phe[8] and Met[11] were investigated since these residues provide the most important contributions to the activity of SP.

Table 2    Antagonist activities of (S)-spirolactam
           substituted compounds

                    7    8    [9,   10]     11
              X-Phe-Phe-(S)-spirolactam-Y.NH$_2$

| Comp. No. | X | Y | pK$_B$ values NK-1 | NK-2 |
|-----------|---|---|------|------|
| (9)  | Ava | Met | 5.6 | <4 |
| (28) | Ava | D-Met | 5.7 | <4.5 |
| (29) | Ava | Phe | 6.3 | 5.4 |
| (30) | Ava | Cha | 6.4 | 5.9 |
| (31) | Ava | HPhe | 6.6 | 5.8 |
| (32) | Ava | Trp | 6.6 | 5.3 |
| (33) | Ava | NH.CH$_3$ | <4.5 | <4.5 |
| (34) | Pro-Gln-Gln | Met | 5.8 | <4.5 |
| (35) | Pro-Lys-Pro-Gln-Gln | Met | 6.4 | <4.5 |
| (36) | Arg-Pro-Lys-Pro-Gln-Gln | Met | 6.9 | 4.8 |
| (37) | Arg-Pro-Lys-Pro-Gln-Gln | Trp | 7.7 | 4.8 |

pK$_B$ values (negative logarithm of apparent
dissociation constant of the antagonist) were determined
from the rightward shift of the agonist dose-response
curve [Substance P methylester in GPI (NK-1) and
neurokinin A in RC (NK-2)] under the conditions described
in Table 1.

Compounds (34)-(37) were synthesized using the Fmoc -
polyamide strategy.[14]

Abbreviations:   Ava, δ-aminovaleryl; Cha, (L)-cyclo-
hexylalanyl; HPhe, (L)-homophenylalanyl; [(S)-2-amino-4-
phenylbutanoyl].

Surprisingly, all position 7 and 8 substitutions either
completely abolished activity (Ala[7], Trp[7], D-Phe[7], D-Trp[7],
Cha[7], Ala[8], Val[8]) or resulted in substantial decreases
(D-Phe[8], pK$_B$ 5.0; Cha[8], pK$_B$ 5.3).  In contrast,
considerable structural variation of the C-terminal
residue (Met[11]) was tolerated with a trend towards
increasing activity with increasing size and lipophilicity
[compounds (28)-(32)].  The activity losses caused by Trp
and D-Trp modifications at position 7 in the antagonists
parallels structure-activity trends in the corresponding
agonist series [compounds (11) and (12), Table 1].
However, the beneficial effects of D-Met[11] and Trp[11]

substitutions for antagonist activity contrast with the
marked reductions in agonist activity caused by the same
modifications of Met[11] in [Ava[6]]-SP(6-11) (Table 1).
These structure-activity relationships suggest that
agonists and antagonists adopt different receptor binding
modes with respect to the C-terminal residue, but may
share common binding sub-sites for residues 7 and 8.

The notion of a common N-terminal receptor binding
mode for agonists and antagonists is further supported by
the progressive increases in antagonist potency obtained
by extension of the N-terminal sequence corresponding to
the native SP sequence [compounds (34)-(36)]. Thus, the
full sequence antagonist (36) has 20 fold (1.3 log units)
greater affinity than the original hexapeptide (9), a
factor of similar magnitude to the 40 fold (1.6 log units)
greater inhibitory potency of SP (pIC$_{50}$ 8.4) relative to
[Ava[6]]-SP(6-11) (pIC$_{50}$ 6.8) measured in a $^3$H-SP radio-
ligand binding assay using GPI membranes.

The effective N-terminal extension and C-terminal
modification (Trp[11]) combined additively in compound (37)
to provide the most potent NK-1 antagonist reported to
date (pK$_B$ 7.7). A Schild plot analysis (slope 0.97,
not significantly different from unity) was consistent
with simple competitive antagonism. The high selectivity
of the compound is demonstrated by its extremely low
activity at NK-2 receptors (NK-1:NK-2 ratio ~ 1000) and by
its lack of activity against neurokinin B in rat portal
vein (NK-3) and against non-peptide agonists (acetyl-
choline and histamine) in the GPI. Furthermore, the
spirolactam antagonist (37) shows negligible affinity for
bombesin or CCK receptors in guinea-pig gall bladder.

From molecular modelling and $^1$H-NMR studies we propose the partial conformations (residues 8-10) for the
spirolactam-containing peptides (8) and (9) illustrated in
Figure 5. The (S)-spirolactam-constrained conformation
lies in the major low energy area of $\phi_{10}$, $\psi_{10}$ space calculated for the model dipeptide analogue, CH$_3$CO-(S)-
spirolactam-NHCH$_3$, and closely resembles a classical type
II' β-turn.[23],[24] An intramolecular hydrogen bond (Met[11]
NH to Phe[8]CO) is evidenced by the low value of the temperature coefficient of the Met[11] NH resonance (Table 3).[25]
The structure is also consistent with the n.O.e. observed
between the Met[11] NH and a γ-lactam ring proton (2.8A in
models). In contrast, for the (R)-spirolactam containing
peptide there was no NMR evidence for a well-defined
structure in solution, but energy calculations favoured
the extended conformation as shown (Figure 5b).

Table 3   Temperature coefficients (p.p.m. x $10^{-3}$/K) for NH resonances in DMSO-$d_6$ (298-338°K).

|        | (9)  | (8)  |
|--------|------|------|
| Phe[7] | -5.5 | -4.3 |
| Phe[8] | -7.3 | -7.5 |
| Met[11]| -1.0 | -4.8 |

One dimensional proton spectra were assigned from 2D COSY and ROESY[26] spectra).

Based on these studies and on the structure-activity relationships described earlier, we propose a receptor binding model for the antagonists (36) and (37) in which the N-terminal (1-8) residues bind in a structurally analogous manner to SP itself, whereas the C-terminal (10 and 11) residues are compelled by the (S)-spirolactam β-turn constraint to adopt a non-activating receptor binding mode. Thus an important role of the constraint may be to prevent access of the C-terminal residue to a critical subsite for receptor activation. But in the resulting antagonist binding mode, the C-terminal moiety is still able to contribute effectively to receptor affinity, as demonstrated by the inactive deletion analogue (33). However, differences in the C-terminal recognition sites for agonists and antagonists are clearly revealed by dissimilar C-terminal structure-activity relationships for the two series.

The recognition of distinct conformations of a peptide ligand by the same receptor in its activated and non-activated states may be a widespread phenomenon which can be exploited for the discovery of antagonists. Our studies have identified a method of distinguishing the conformational requirements for receptor recognition of peptide agonists and antagonists. This approach has provided the most potent and selective neurokinin NK-1 antagonist so far described, and we are currently investigating further applications.

## Acknowledgements

We thank Dr M.M. Hann and Dr. A.P. Tonge for molecular modelling studies and Dr. B. Carter for NMR measurements.

REFERENCES

1.  D. Regoli, Trends Pharmacol.Sci., 1985, 7, 481.

2.  M. Rodriguez, J-P. Bali, R. Magous, B. Castro and
    J. Martinez, Int.J.Peptide Protein Res., 1986, 27,
    293.

3.  D.H. Coy et al., J.Biol.Chem., 1988, 263, 5056.

4.  H. Schulz and V. Du Vigneaud, J.Med.Chem., 1966, 9,
    647.

5.  G. Skala et al., Science, 1984, 226, 443.

6.  M.M. Chang, S.E. Leeman and H.D. Niall, Nature
    (London), New Biol., 1971, 232, 86.

7.  X.-Y. Hua, et al., Regulatory Peptides, 1985, 13, 1.

8.  K. Folkers et al., Acta Chem.Scand., 1982, 36,
    389.

9.  D. Regoli, E. Escher and J. Mizrahi, Pharmacology,
    1984, 28, 301.

10. L. Zhang, S. Mantey, R.T. Jensen and J.D. Gardner,
    Biochim.Biophys.Acta, 1988, 972, 37.

11. A.T. McKnight, J.J. Maguire, B.J. Williams,
    A.C. Foster, R. Tridgett and L.L. Iversen,
    Regulatory Peptides, 1988, 22, 127.

12. M.A. Cascieri et al., Molec.Pharmac., 1986, 29,
    34.

13. J.R. Brown et al., 'Tachykinin Antagonists' ed.
    R. Hakanson and F. Sundler, pp. 305-312, Elsevier,
    Amsterdam, 1985.

14. E. Atherton, C.J. Logan and R.C. Sheppard,
    J.Chem.Soc.Perkin Trans.I, 1981, 538.

15. R. Laufer, C. Gilon, M. Chorev and Z. Selinger,
    J.Med.Chem., 1986, 29, 1284.

16. C.-M. Lee, N.J. Campbell, B.J. Williams and
    L.L. Iversen, Eur.J.Pharmac., 1986, 130, 209.

17. G.E. Schulz and R.H. Schirmer, 'Principles of Protein Structure,' Springer-Verlag, New York, 1979.

18. F.H. Allen, O. Kennard and R. Taylor, Acc.Chem.Res., 1983, 16, 146.

19. R.M. Freidinger, D.S. Perlow and D.F. Veber, J.Org.Chem., 1982, 47, 104.

20. D. Seebach, M. Boes, R. Naef and W.B. Schweizer, J.Am.Chem.Soc., 1983, 105, 5390.

21. L. Kisfaludy et al., Tet.Letters, 1974, 1785.

22. M.G. Hinds, N.G.J. Richards and J.A. Robinson, J.Chem.Soc., Chem.Commun., 1988, 1447.

23. P.Y. Chou and G.D. Fasman, J.Molec.Biol., 1977, 115, 135.

24. P.N. Lewis, F.A. Momany and H.A. Sheraga, Biochim. Biophys.Acta, 1973, 303, 211.

25. K.D. Kopple, M. Ohnishi and A.Go, J.Am.Chem.Soc., 1969, 91, 4264.

26. C. Griesinger and R.R. Ernst, J.Magnetic Resonance, 1987, 75, 261.

# The Discovery and Design of Substrate-based Proteinase Inhibitors—Problems and Lessons from the Development of Renin Inhibitors as Potential Antihypertensive Drugs

By R. J. Arrowsmith*, C. J. Harris*, D. E. Davies*, J. A. Morton*, J. G. Dann†, and J. N. Champness‡

DEPARTMENT OF MEDICINAL CHEMISTRY*, DEPARTMENT OF BIOCHEMICAL SCIENCE†, DEPARTMENT OF PHYSICAL CHEMISTRY‡, WELLCOME RESEARCH LABS., LANGLEY PARK, BECKENHAM, KENT BR3 3BS, UK

Introduction: The Renin-Angiotensin System

The renin-angiotensin system (RAS) is intimately involved in blood pressure homeostasis[1]. The potent octapeptide vasoconstrictor angiotensin II (AII) ·is produced from its precursor, angiotensinogen, by the action of two proteinases. The first of these, renin, is a highly specific aspartyl proteinase secreted by the juxta glomerular cells of the kidney. Renin cleaves angiotensinogen at the N-terminus, releasing the decapeptide angiotensin I (AI), as shown in Figure 1. The C-terminal dipeptide of AI is then cleaved by the metalloproteinase angiotensin converting enzyme (ACE) to release the pressor peptide AII. Intervention in this cascade has proved therapeutically (and commercially) successful, with the development of ACE inhibitors as effective antihypertensive agents[2]. As a result, much interest has been focussed on the potential of renin inhibitors as an alternative point of intervention in the RAS. Indeed, it is possible that renin inhibitors may have advantages over ACE inhibitors, since the reaction catalysed by renin is the rate determining step in the cascade, and because ACE is known to hydrolyse a number of substrates, whereas renin is considered to be monofunctional[3].

In our efforts to design a useful renin inhibitor, we considered it a central issue that a candidate drug should be capable of competing with existing therapies, particularly ACE inhibitors. This requires a small,

potent orally active inhibitor lacking the need for
complex peptide synthesis and purification.

$Asp^1.Arg^2.Val^3.Tyr^4.Ile^5.His^6.Pro^7.Phe^8.His^9.Leu^{10}.Val^{11}.Ile^{12}.His^{13}.Asn^{14}$. - PROTEIN

ANGIOTENSINOGEN

ANGIOTENSIN I

ANGIOTENSIN II

**Figure 1**   Enzymatic Cleavage of Human Angiotensinogen

**Substrate based inhibitors.**

Renin cleaves the $Leu^{10}$-$Val^{11}$ amide bond of
angiotensinogen (Figure 1). Various non-hydrolysable
isosteres of this bond (Figure 2) have been designed and
incorporated into modified peptide sequences to provide
potent inhibitors of human renin.

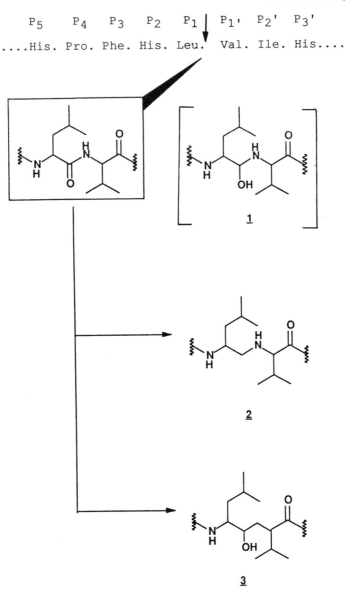

**Figure 2**   Some non-hydrolysable isosteres of the Leu[10]-Val[11] amide bond.

Two examples are the reduced bond (2)[4] and hydroxyethylene (3)[5] isosteres employed in the inhibitors H-142 and H-261 respectively (Table 1). The naturally occurring amino acid statine has also been widely used in similar sequences[6] (*e.g.* See Table 1). Statine is considered to be a transition state analogue, and to interact with the two catalytically essential aspartic acid residues of renin (Asp [215] and Asp [32]) as shown schematically in Figure 3.

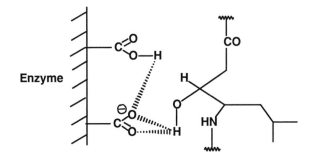

Enzyme

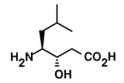

$H_2N$ ⋮ $CO_2H$
$OH$

**Statine**

<u>Figure 3</u>   Statine Inhibitors

Table 1

IC$_{50}$ (μM, human renin)

| | |
|---|---|
| H-142 | H.Pro.His.Pro.Phe.His.Leu.$\underline{R}$ Val.Ile.His.Lys.OH | 0.01 |
| H-261 | Boc.His.Pro.Phe.His.Leu.$\underline{H.E}$.Val.Ile.His.OH | 0.0007 |
| Merck | Iva.His.Pro.Phe.His.Sta - Leu.Phe.NH$_2$ | 0.05 |
| Merck | Iva.His.Pro.Phe.His.Epi.Sta - Leu.Phe.NH$_2$ | 14.0 |

We proposed that replacing the hydroxyl group of statine with an ammonium group may provide additional binding to the enzyme, as a result of electrostatic interaction with the catalytic aspartic acid residues in the active site.

Although the $\underline{S}$-amino enantiomers of sequences containing this isostere (Asta, aminostatine) showed no significant gain in potency compared with analogous statinoid sequences, the $\underline{R}$-amino enantiomers were considerably more potent than the corresponding statinoid compounds. This has been rationalised in terms of a contribution to the overall binding from the proposed electrostatic interaction[7].

It should be noted that the size and peptidic nature of the potent inhibitors listed in Table 1 make them unlikely drug candidates in terms of the criteria discussed earlier. Potency in analogues containing these isosteres falls dramatically with decreasing length of the peptidic chain.

The Amino-alcohol isostere

We have investigated a range of sequences that incorporate the novel amino-alcohol isostere (<u>4</u>).

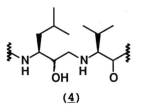

(<u>4</u>)

This is formally derived from the unstable hemi-aminal (1) by the insertion of an extra methylene group

into the backbone of the inhibitor. Synthetic details for
inhibitors based on this isostere have been described
previously[8]. As expected, potent inhibition is observed
in octapeptide analogues[9] and as with statine or
hydroxyethylene inhibitors[10], there is a small but
significant preference for enantiomers having S-hydroxyl
stereochemistry (BW 807C and 808C, Table 2). In contrast
to other isosteres however, amino-alcohols also provide
reasonably potent inhibitors in the severely-truncated
tetrapeptide sequence, as illustrated, for example, by
BW684C and BW625C, Table 2. Furthermore, in these shorter
sequences, the stereochemical preference is reversed *i.e.*,
compounds with R-hydroxyl stereochemistry are now more
potent (BW 624/625C), indicating different modes of
binding for the tetrapeptide versus octapeptide analogues.
Blockade of the N-terminus of tetrapeptide inhibitors is
associated with a marked drop in potency. This led us to
the hypothesis that the N-terminal ammonium group may be
salt-bridging to the catalytic aspartic acid residues, in
a manner similar to that proposed for Asta containing
compounds[7].

## Computer Graphics Studies

Using a model of human renin based upon the known
structure of endothiapepsin, an aspartic protease
homologous to renin[11], we attempted to fit the inhibitors
BW624C and BW625C to the enzyme with an N-terminal
interaction with the active site as described above. Only
the structure having the R-hydroxyl configuration (BW625C)
could be accomodated, with the hydroxyl group itself
making a potential hydrogen-bonded interaction with one of
the so-called flap residues, Ser [76]. The hydroxyl group
of the S-epimer (BW624C) appeared to make an unfavourable
interaction with the aromatic ring of Tyr [75]. Encouraged
by the correlation between modelling and observed
activity, we used this model to design further
tetrapeptide analogues based upon the proposed salt-
bridging mode of binding. According to this model,
occupancy of the $S_1$ subsite by the isobutyl side chain at
$P_1$ in BW625C was poor. The best fit in this position was
achieved with the 3-phenylpropyl side-chain, which
appeared to exploit the potential hydrophobic binding more
fully. Indeed, the resulting inhibitor provided an extra
order of magnitude in potency *in vitro* (BW625C vs. BW128C,
Table 2). Further refinement of the structure using a
combination of computer modelling and classical SAR
techniques resulted in the potent inhibitor BW586C.

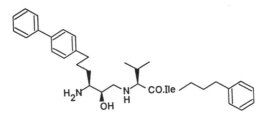

BW586C                    $IC_{50}$ 0.04 $\mu$M

## Table 2

                                                                          $IC_{50}$ $\mu$M

| BW807C | H.His.Pro.Phe.His.Leu | AA Val.Ile.Phe.OMe | (S) | 0.07 |
|--------|----------------------|--------------------|-----|------|
| BW808C | H.His.Pro.Phe.His.Leu | AA Val.Ile.Phe.OMe | (R) | 0.19 |
| BW684C | H.Sta | - Val.Ile.Phe.OMe | | 1105 |
| BW624C | H.Leu | AA Val.Ile.Phe.OMe | (S) | 111 |
| BW625C | H.Leu | AA Val.Ile.Phe.OMe | (R) | 6.0 |
| BW128C | HPpg | AA Val.Ile.Phe.OMe | (R) | 0.4 |
| BW586C | H.(pPh)Ppg | AA Val.Ile.NH∿∿Ph | (R) | 0.04 |

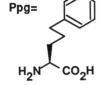

Ppg=

H₂N  CO₂H

**Phenylpropylglycine**

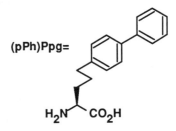

(pPh)Ppg=

H₂N  CO₂H

**R-Phenylphenylpropylglycine**

## X-Ray Studies

Both BW624C and BW625C have been co-crystallised with
endothiapepsin[11] in collaboration with Prof. Tom Blundell
and his co-workers at Birkbeck College, University of
London.  Preliminary results on the BW624C complex have

been reported elsewhere[12]. In summary, the inhibitor binds as predicted from $S_1$ to $S_3'$, but the main interactions with the enzyme are essentially the same as those seen in complexes with reduced and hydroxyethylene-type inhibitors. Therefore, it is the S-hydroxyl group, not the terminal ammonium group, which interacts with the active site aspartyl residues. The ammonium group may participate in a weak interaction with the carbonyl group of Gly[217].

The more potent R-hydroxyl diastereoisomer, BW625C, also binds mainly in the $S_1$-$S_3'$ subsites, but, in this case, there is evidence of disorder in the crystal, with the probability that more than one distinct enzyme-inhibitor complex is present. Thus, at least as far as endothiapepsin is concerned, our simple model of binding is unproven at this stage, but has nevertheless proved to be useful predictively.

It should be noted that there are interesting discrepancies in binding to renin and endothiapepsin for this class of inhibitor which are evident from inhibition measurements (Table 3). Both BW624C and BW625C inhibit endothiapepsin more strongly than they do renin, but again it is the R-hydroxyl epimer which is the more potent. Removal of the hydroxyl function (BW572C, Table 3) is detrimental to renin inhibition, indicating that it provides a favourable interaction with the enzyme in this case.

Table 3

|  |  | Endothia $IC_{50}$ µM | Renin $IC_{50}$ µM |
|---|---|---|---|
| BW624C | H.Leu AA Val.Ile.Phe.OMe (S) | 0.96 | 110 |
| BW625C | H.Leu AA Val.Ile.Phe.OMe (R) | 0.04 | 6 |
| BW572C | H.Leu DES-OH Val.Ile.Phe.OMe | 0.08 | 156 |

In contrast, BW572C and BW625C are equipotent against endothiapepsin, which implies that the hydroxyl group makes no significant contribution to binding, and may even make a negative contribution when present in the S-configuration (BW624C, Table 3). These results suggest that tetrapeptide amino-alcohol inhibitors bind in fundamentally different ways to renin and endothiapepsin and the relevance of the X-ray data in terms of the design

of renin inhibitors must be in doubt.  However, the
endothiapepsin studies have illustrated that many
alternative binding modes may be available, particularly
to small inhibitors interacting with each individual
protease.  This is obviously of concern from the point of
view of drug design.

Non-Prime Amino alcohols.

    We have investigated the applicability of the amino-
alcohol isostere in inhibitors that bind in the $P_3-P_1$
subsites, as well as $P_1-P_3'$ subsites as discussed above.
In this series, there is also the possibility of
electrostatic interaction between the amino-alcohol amine
function, and the catalytic aspartyl residues (Figure 4)

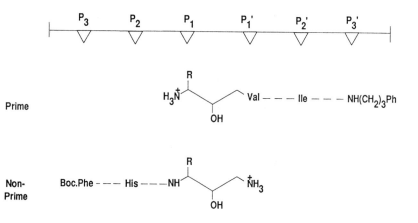

Figure 4

    Detailed synthetic routes to these inhibitors will be
published elsewhere.  Relevant SAR from the non-prime
series is given in Table 4.

$IC_{50}$ µM

| | | | |
|---|---|---|---|
| BW115C | Boc.Phe.Ala.Leu $\overline{AA}$ H | (S) | 300 |
| BW418C | | (R) | 16 |
| | | | |
| BW958C | Boc.Phe.Ala.Ppg $\overline{AA}$ H | (MP) | 70 |
| BW957C | | (LP) | 48 |
| | | | |
| BW747C | Boc.Phe.His.Leu $\overline{AA}$ H | (S) | 90 |
| BW746C | | (R) | 5 |
| | | | |
| BW34C | Boc.Phe.Ala.Cha $\overline{AA}$ H | (S) | 97 |
| BW33C | | (R) | 1.4 |
| | | | |
| BW490C | Boc.Phe.His.Cha $\overline{AA}$ H | (S) | 0.25 |
| BW489C | | (R) | 0.76 |
| | | | |
| | | | |
| BW647C | Boc.Phe.His.Cha[CH(OH)CH$_2$N$_3$] | (S) | 0.07 |
| BW646C | | (R) | 2.5 |

Table 4

Incorporation of the $P_1$-Leu amino-alcohol isostere into $P_3$-$P_1$ analogues gives inhibitors of relatively low potency, with $P_2$-His (BW746C), the natural $P_2$ residue, being preferred over $P_2$-Ala (BW418C) as expected. In contrast to the primeside inhibitors previously discussed, extended lipophilic side chains at $P_1$ are not favoured (BW957/958C), and the well-known cyclohexylmethyl (Cha) $P_1$ substituent appears optimal, providing sub-micromolar affinities (BW 489/490C). In these more potent inhibitors, however, it is surprising to see that the common R-hydroxyl stereochemical preference for amino alcohol-containing inhibitors has reversed, with a slight preference for the S-hydroxyl stereochemistry becoming evident.

Replacement of the strongly basic C-terminal primary amine group with an azido group (BW 646/647C) results in a marked increase in affinity for the enzyme, and this increase is associated with the S-hydroxyl stereochemistry. Thus it appears that basic groups are detrimental to affinity in this series.

References

1.   M.J. Peach, Physiol. Rev., 1977, 57, 313.

2.   M.A. Ondetti and D.W. Cushman, Annu. Rev. Biochem.,
     1982, 51, 283.

3.   B.Foltmann and V.B. Pedersen, in 'Acid Proteases,
     Structure, Function and Biology', J. Tang, Ed.,
     Plenum, New York, 1977, 3.

4.   M.Szelke, B. Leckie, A. Hallet, D.M. Jones, J.
     Sueires, B. Atrash and A.F. Lever, Nature, 1982, 299,
     555.

5.   M. Szelke, in 'Aspartic Proteases and their
     Inhibitors', V. Kostka, Ed., de Gruyter, Berlin,
     1985, 421.

6.   J. Boger, N.S. Lohr, E.H. Ulm, M.Poe, E.H. Blaine,
     G.M. Fanelli, T.Y. Lin, L.S. Payne, T.W. Schorn, B.I.
     LaMont, T.C. Vassil, I.I. Stabilito, D.F. Veber, D.H.
     Rich and A.S. Bopari, Nature, 1983, 303, 81.

7.   R.J. Arrowsmith, K.R. Carter, J.G. Dann, D.E. Davies,
     C.J. Harris, J.A. Morton, P. Lister, J.A. Robinson
     and D.J. Williams, J. Chem. Soc., Chem. Commun.,
     1986, 755.

8.   R.J. Arrowsmith, D.E. Davies, Y.C. Fogden, C.J.
     Harris and C. Thomson, Tet. Lett., 1987, 28 (45),
     5569.

9.   J.G. Dann, D.K. Stammers, C.J. Harris, R.J.
     Arrowsmith, D.E. Davies, G.W. Hardy and J.A. Morton,
     Biochem. Biophys. Res. Comm., 1986, 71.

10.  D.H. Rich and E.T.O. Sun, J. Med. Chem., 1980, 23,
     27.

11.  B.L. Sibanda, T.L. Blundell, P.M. Hobart, M.
     Fogliano, J.S. Bindra, B.W. Dominy and J.M. Chirgwin,
     Febs Lett., 1984, 174, 102.

12.  J.B. Cooper, S.I. Foundling, T.L. Blundell, R.J.
     Arrowsmith, C.J. Harris and J.N. Champness, in
     'Topics in Medicinal Chemistry', 4th RSC-SCI Med.
     Chem. Symposium, P.R. Leeming, Ed., RSC, 1988, 308.

# Recent Developments in the Solid Phase Synthesis of Peptide Immunogens

By T. Johnson and R. C. Sheppard

MRC LABORATORY OF MOLECULAR BIOLOGY, HILLS ROAD, CAMBRIDGE CB2 2QH, UK

Peptides and proteins are central figures in biomolecular recognition. Enzyme-substrate and hormone-receptor interactions, and the workings of the immune system are examples which come readily to mind. In the last case, short peptides can often stimulate the production of antibodies which recognise not only the original immunogen but also proteins containing the same amino acid sequence. This discovery has opened a whole new field of research. Antibodies to synthetic peptides may now be used to identify, locate in biological structures, and isolate proteins whose existence were conjectured only from gene structures. Peptide synthesis has thus become an important tool in molecular biology.

This paper reviews some of the recent developments in peptide synthesis which have helped to make these substantial sized molecules readily accessible. Most peptide synthesis is currently carried out for raising antibodies, and some new developments are indicated which may facilitate the generation of peptides in an immunogenic form directly suitable for injection into experimental animals.

Nearly all peptide synthesis of sizeable targets (say above ten amino acid residues) is now necessarily carried out following Merrifield's solid phase principle.[1] This is an accelerated procedure in which the peptide chain is assembled on a solid (usually gel phase) insoluble support by a series of stepwise amino acid coupling and deprotection reactions. It is rapid, simple to carry out, and remarkably effective, though it flouts some of the basic tenets of classical organic synthesis. In particular, it provides no opportunity for the isolation, purification, and characterisation of intermediates during the synthesis. Products of incomplete reactions or of side reactions may accumulate on the solid support and contaminate the desired product. In principle, therefore, solid phase synthesis can be completely satisfactory only if:

(1) All coupling and deprotection reactions are strictly quantitative; and

(2) There are no side reactions involving, for example, reactive amino acid side chains; or

(3) Impurities arising through inadequacies in (1) or (2) can be removed (and shown to be removed).

The organic chemist will recognise that with molecules as large, complex, and sensitive as peptides, none of these goals is likely to be completely achievable. For consideration (1) above, the simple arithmetic of repeated consecutive reactions is awsome, and yields in excess of 99% for each must be aimed for if useful products are to be obtained. New methods based upon the solid phase principle can at best improve the closeness to which these ideals may be approached. Regrettably, purity of the final product is not easily judged in absolute terms, and is often assessed by its suitability for the proposed application.

Our efforts to improve the efficiency and reliability of the solid phase procedure led to a complete reappraisal of both reaction conditions and protecting group combinations.[2] Special attention was paid initially to ensuring full solvation of the peptide polymer-complex by solvents likely to be kinetically suitable for both the acylation and deprotection reactions. Experience from solution chemistry suggested that dipolar aprotic solvents of the dimethylformamide type were most likely to fulfil this last role for the acylation step. Dimethylformamide itself is one of the better solvents for protected peptides. To ensure solvation of the solid support, a new polar polydimethylacrylamide gel (1) was synthesised.[3] As expected, this polymer swelled dramatically in contact with dimethylformamide, one gram of the dry polymer occupying a volume of 20 ml in contact with the solvent. Thus the polymer

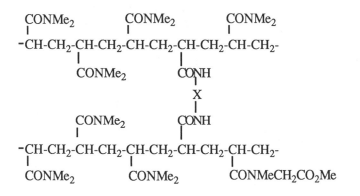

$(1, X=-CH_2CH_2-;\ 2, X=-CH_2CH_2-OCMe_2O-CH_2CH_2-)$

itself contributes rather little to the reaction medium and being structurally related to the solvent, may be considered as 'transparent' to the reactants. The methyl ester group introduced by inclusion of acryloyl-sarcosine in the monomer mixture provides the attachment point for the growing peptide chain. A structural variant (2) of the polymer[4] can be transformed into a soluble form enabling further manipulation in solution after completion of the solid phase assembly (see below). Both resins have been prepared in a novel physically supported form by polymerising the monomer mixture within the pores of rigid macroporous inorganic particles. This provides a particulate support suitable for peptide synthesis under pumped flow conditions without compression and the generation of high pressures.[5] Diffusion of solvent and reactants into and out of the gel phase is not impaired. The design of this polymeric support has enabled construction of a new generation of continuous flow peptide synthesisers with substantial advantages in terms of ease of operation, mechanical simplicity, and especially provision of analytical control.[5]

The choice of protecting groups is crucial in peptide synthesis. Temporary protection of the terminal amino group is required together with more permanent protection of side chain functional groups. The overiding considerations for α-amino protection is complete masking of the reactivity of the amino group and ease of repeated removal under mild reaction conditions which do not affect protecting groups elsewhere. We selected the base-labile fluorenylmethoxycarbonyl (Fmoc) group (3) for the α-amino function.[6,7] It is cleaved rapidly and completely by piperidine in dimethylformamide under very mild conditions. Its strong uv absorption offered possibilities for spectrometric monitoring of reactant concentration in the flowing stream for both acylation and deprotection steps. Choice of a base-labile α-amino protecting group enabled acid-labile t-butyl or *p*-alkoxybenzyl-based derivatives to be used for side chain and carboxy-terminus protection. The latter usually constitutes the linkage of the peptide to the resin support. There is complete differentiation between the permanent and temporary protecting groups, and each type may be cleaved completely under mild reaction conditions in the presence of the other.

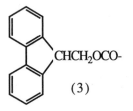

(3)

For the acylation step, the requirements are for high reactivity without side reactions. Symmetric anhydrides of Fmoc-amino acids (4) proved very satisfactory[7] but have now been largely replaced by the more convenient pentafluorophenyl esters (5).[8,9] In most cases these are stable crystalline solids which react smoothly with amino groups and offer little scope for side reactions. Their reactivity can be enhanced if necessary by addition of catalyst 1-hydroxybenzotriazole. More recently we have also introduced esters of Fmoc-amino acids with 3,4-dihydro-3-hydroxy-4-oxo-1,2,3-benzotriazine (6).[10,11] These have reactivity comparable to symmetric anhydrides, but their main interest lies in the opportunity they present for following the course of the acylation reaction on the solid support. As ammonolysis takes place, the weakly acidic 3-hydroxybenzotriazine liberated from the ester forms an ion pair with residual amino groups on the resin.[11] The ionised form of the hydroxybenzotriazine is bright yellow, and the resin (but not the recirculating solution) immediately becomes strongly coloured when the reactant is introduced and then fades as the amino groups are consumed. A simple photometric device for measuring this resin colour has been constructed, enabling peptide synthesis to be carried out with feedback analytical control in a fully automated manner.[12]

(Fmoc-NHCHRCO)$_2$O   (4)      Fmoc-NHCHRCO-OC$_6$F$_5$   (5)

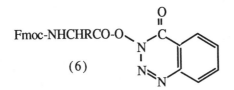

(6)

These and other chemical features have now been integrated into a highly developed system for solid phase synthesis under continuous flow conditions. Commercial equipment is available for carrying out the various chemical operations sequentially under manual or computer control. Such synthesisers are frequently operated in a biological rather than chemical environment and under circumstances where careful chemical characterisation of the products is not always possible. Continual reassurance is therefore desirable that the necessary very high chemical efficiencies can in fact be achieved. Although each example is an individual case, I would like to describe briefly one recent synthesis of a thirty residue sequence where the product has been examined very closely.

-Pro-Thr-Ser-Pro-Ser-Tyr-Ser-     (7)

H-(Pro-Thr-Ser-Pro-Ser-Tyr-Ser)-Pro-Gly-OH     (8)

The major sub-unit of RNA polymerase II contains the heptapeptide unit (7) repeated many times.[13] For studies on the transcription process, the synthetic peptides (8) with n=2, 3, and 4 were required. For the 30 residue peptide containing four heptapeptide units, the sixty consecutive acylation and deprotection reactions would require an average efficiency of 99% to yield a crude product of 55% purity. With 99.5% efficiency, the initial purity would be about 74%.* In the event, a straightforward stepwise assembly using Fmoc-amino acid pentafluorophenyl and dihydrooxobenzotriazinyl esters on polydimethylacrylamide resin in a Pharmacia-LKB Biolynx peptide synthesiser gave an unpurified product with the hplc profile shown in figure 1. Such chromatograms are reassuring, but hardly constitute rigorous proof of purity at this molecular weight level. A resin sample at the 29 residue stage was also cleaved, as was another containing 31 residues prepared from the main peptide-resin by a further cycle of acylation and deprotection. Both crude peptides gave chromatograms similar to that in figure 1, and although the elution times were slightly

---

* In considering these figures, it should be remembered that most of the impurities caused by incomplete reaction will consist of many hard to separate sequences deficient in just one or two amino acid residues.

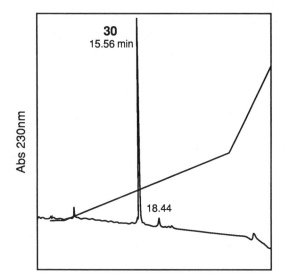

Figure 1. Hplc of total crude 30 residue peptide (8, n=4). Conditions: elutant A, 0.1% aq trifluoroacetic acid; B, 90% acetonitrile, 10% A. Gradient 10-40% B over 25 minutes.

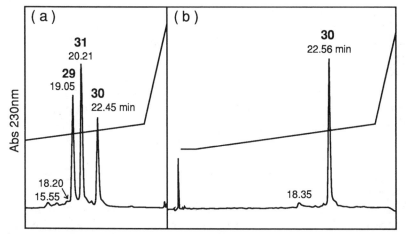

Figure 2. (a) Hplc resolution of a mixture of synthetic 29, 30, and 31 residue peptides. (b) Total crude 30 residue peptide. Conditions: A, 5mM aq sodium heptane sulphonate; B, 90% acetonitrile, 10% A. Gradient 12-22% B over 25 minutes.

different, the three peptides co-eluted as a single, slightly broader peak. Routine hplc may therefore be quite inadequate in assessing the purity of peptides of this length. Only if the analytical tool employed is demonstrably capable of detecting likely contaminants can it be interpreted with confidence. A careful study using several elutant combinations and gradients produced a system in which a mixture of the 29, 30, and 31 residue peptides were clearly separated (figure 2,a). Under these conditions, the target 30 residue product gave the profile of figure 2,b and the adjacent 29 and 31 residue homologues were again similar. Clearly both acylation and deprotection reactions were proceeding with very high efficiency even at this chain length.

Fortified with this experience, we were tempted to take one further step towards the simplification of peptide synthesis for generating antibodies. Most small peptides are not intrinsically immunogenic and require to be attached to a high molecular weight carrier molecule before introduction into the experimental animal. Proteins are most commonly used as carriers, though other non-proteinaceous polymers, notably polylysine, have also been used from time to time. The immune system is clearly incredibly complex and details of its workings are only beginning to emerge. It is probable that the protein carriers themselves contribute peptide sequences or fragments which aid antibody formation by the attached immunogen, but this is not essential. It was therefore very tempting to consider whether a polymer support could be made which was suitable both for solid phase peptide synthesis and directly for raising antibodies to the attached peptide. Detachment of the synthetic peptide from the synthesis support and re-attachment to a second carrier would not be required. This last process is commonly carried out using reagents such as carbodiimides which may cause significant structural modification to both the peptide and its carrier. The new system envisaged could therefore produce a more structurally defined system. It could provide an economic route to the large scale production of synthetic peptide vaccines.

Early experiments suggested that the gel support (1) was unlikely to fulfil this role. A more promising approach was to design a modification of (1) which retained the insoluble gel characteristics required for synthesis in an organic medium, but which could then be *solubilised* to provide a peptide-carrier conjugate in homogeneous aqueous solution. The solubilisation process should also remove all protecting groups from the peptide component, but not of course cleave the peptide-polymer linkage.

This proved relatively simple to achieve.[4] Insolubility of the synthetic polymer (1) is conveyed by the cross linking of linear chains. When the simple bisacryloyl ethylene diamine monomer was replaced in the polymerisation mixture by an analogous acetal or ketal structure, the resulting insoluble polymer (2) was readily cleaved by acid to a soluble, high molecular weight linear polymer.

Some further modification was required to ensure suitability for peptide synthesis. Conventionally, the methyl ester groups in (1) are extended to amino functions by reaction with excess ethylene diamine as a first step in synthesis. For analytical purposes, an internal reference amino acid is then coupled to this amino group, followed by a specifically cleavable benzyl alcohol linkage agent for attaching the peptide chain. It became evident with the new polymer (2) that the ethylene diamine step was introducing a small degree of additional cross linking preventing complete solubility of the acid-treated product. The acryloylsarcosine methyl ester monomer used to introduce the initial methyl ester was therefore replaced by the acryloyl derivative of 1-amino-6-acetoxyhexane. After cleavage of the acetyl protecting group, this provided a polymer related to (2) functionalised with simple hydroxyl groups for initiating peptide synthesis. For characterisation purposes, peptides synthesised on this polymer may be detached after solubilisation by ammonolysis or alkaline hydrolysis.

This new support for peptide synthesis and immunology is giving encouraging results. Sequences related to substance P and substance K (neurokinin A),[4] and more recently several short part sequences of the protein endoplasmin have been synthesised on the polymer. All were immunogenic. Antibodies in the endoplasmin series could be isolated using the same soluble polymer-peptide adsorbed on nitrocellulose film.

Further development of this system may provide a particularly valuable tool in immunochemistry. 'Helper' peptide sequences similar to those derived from protein carriers may be co-synthesised on the same polymer in addition to the desired immunogen. These may increase the immunogenicity by stimulating T and B-cell activity, and facilitate investigation of the biological role of such sequences. It is appreciated that the synthetic peptide-polymer represents a further departure from the rigour of classical chemical synthesis, and care will be needed to ensure the fidelity of the peptide sequence in each case. The evident efficiency of the basic Fmoc-polyamide solid phase technique outlined above is encouraging in this respect.

We are grateful to Dr Gordon Koch for permission to refer to his recent results in the endoplasmin series.

REFERENCES

1.  R.B. Merrifield, *J. Amer. Chem. Soc.*, 1963, **85**, 2149

2.  R.C. Sheppard, *Science Tools*, 1986 **33**, 9

3.  R. Arshady, E Atherton, D.L.J. Clive, and R.C. Sheppard, *J. Chem. Soc., Perkin Trans. 1* , 1981, 529

4.  P. Goddard, J. McMurray, R.C. Sheppard, and P. Emson, *J. Chem. Soc., Chem. Comm.*, 1988, 1025

5.  R.C. Sheppard, *Chem. Brit.* 1983, 402

6. L.A. Carpino and G.Y. Han, *J. Org. Chem.*, 1972, **37**, 3404

7. E. Atherton, C.J. Logan, and R.C. Sheppard, *J. Chem. Soc.*, *Perkin Trans. 1*, 1981, 538

8. L. Kisfaludy and I. Schon, *Synthesis*, 1983, 325; I. Schon and L. Kisfaludy, *Synthesis*, 1986, 303

9. E. Atherton, L.R. Cameron, and R.C. Sheppard, *Tetrahedron*, 1988, **44**, 843

10. W. Konig and R. Geiger, *Chem. Berichte*, 1970, **103**, 2024

11. E. Atherton, J.L. Holder, M. Meldal, R.C. Sheppard, and R.M. Valerio, *J. Chem. Soc., Perkin Trans. 1*, 1988, 2887

12. L.R. Cameron, J.L. Holder, M. Meldal, and R.C. Sheppard, *J. Chem. Soc., Perkin Trans. 1*, 1988, 2895

13. J.L. Corden, D.L. Cadena, J.M. Ahearn, Jr, and M.E. Dahmus, *Proc. Natl. Acad. Sci., USA*, 1985, **82**, 7934

# Probing the Mechanism of Methylaspartase

By Nigel P. Botting and David Gani

DEPARTMENT OF CHEMISTRY, THE UNIVERSITY, SOUTHAMPTON SO9 5NH, UK

## 1 INTRODUCTION

The enzyme β-methylaspartase[1] (3-methylaspartate ammonia-lyase, EC 4.3.1.2) catalyses the reversible α,β-elimination of ammonia from L-*threo*-(2S,3S)-3-methylaspartic acid (1) to give mesaconic acid (2). The enzyme lies on the main catabolic pathway for glutamate in *Clostridium tetanomorphum* and a number of other species[2,3] The Clostridial enzyme, which is the best studied, has been shown to deaminate the L-*erythro*-(2S,3R)-diastereomer of methylaspartic acid[1] as well as (2S)-aspartic acid and a number of 3-alkylhomologues.[4] The enzyme which was reported to possess an (AB)₂ structure[5,6] requires only monovalent and divalent cations for activity.

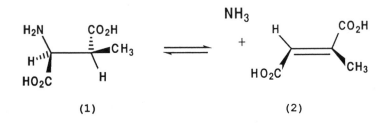

(1)                                    (2)

**Scheme 1**    The Methylaspartase Reaction

Since it was known that the enzyme could be used to cataylse the amination of mesaconic acid to give (2S,3S)-3-methylaspartic acid ( the L-*threo*- isomer) in good yield,[7] we undertook to assess the utility of the enzyme in enantiospecific syntheses of 3-halogeno- and 3-alkyl-L-aspartic acids.[8] We expected that these might serve as irreversible and reversible inhibitors, respectively, for pyridoxal-P dependent enzymes which were under investigation in our laboratory.

Accordingly, a series of 3-substituted fumaric acids were prepared (3, X= H, Me, Et, n-Pr, i-Pr, n-Bu, F, Cl, Br, and I ) and were incubated with the enzyme. Except for n-butyl- and iodo-fumarate, the corresponding 3-substituted aspartic acids were formed. The products derived from the fumaric acids ( X= H, Me, Et, n-Pr, i-Pr, and Cl) could be isolated in good to excellent yield.[9] The stereochemistry of the products (4, X= Cl, Br and Et) were determined by degradation and/or by examination of the [1]H-nmr spectrum of anhydride derivatives. In each case it was evident that during the incubation *anti*- addition of ammonia had occurred to give an L-*threo*- amino acid. (Note the *absolute configurations* are different).[9,10]

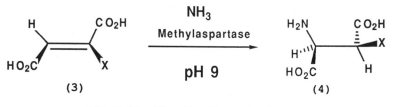

( X= H, Me, Et, n-Pr, i-Pr, and Cl)

Scheme 2 Synthesis of 3-substituted aspartic acids

During these synthetic studies it was noted that the amination rates for the fumaric acids (3, X= H, Me, Cl and Br) were qualitatively similar, however, the reported deamination rate for (2S)-aspartic acid was about 100 times slower than that for the physiological substrate, (2S,3S)-methylaspartic acid.[10] This interesting observation prompted us to investigate, in detail, the rates of reaction for a range of substrates/ products in both the forward (deamination) and reverse reaction direction.


## 2    AMINATION AND    DEAMINATION KINETICS

The kinetic parameters for enzymic amination were determined for a range of substituted fumaric acids (3,X= H, Me, Et, *n*-Pr, *i*-Pr, Cl, and Br) in the presence of 400 mM ammonia at pH 9.0. Initial rates were obtained by measuring the decrease in $A_{240}$ (due to the conjugated double bond of the substrates) spectrophotometrically. The results are shown in Table 1.[11]


**Table 1**    Kinetic  Parameters  for  the  Amination
of Substituted Fumaric acids[a]

| Substrate | $K_m$ (mM) | [b]$k_{cat}$ ($s^{-1}$) |
|---|---|---|
| Fumaric acid | 23.0  ±2.2 | 2170 |
| Mesaconic acid | 1.24 ±0.085 | 1120 |
| Ethylfumaric acid | 1.05 ±0.2 | 730 |
| Chlorofumaric acid | 3.52 ±0.71 | 480 |
| Bromofumaric acid | 2.64 ±0.53 | 534 |
| n-Propylfumaric acid | 2.1  ±1.3 | 5.4 |
| 1-Propylfumaric acid | 5.5  ±3.0 | 6.7 |
| n-Butylfumaric acid | | <0.01 |
| Iodofumaric acid | | <0.01 |


[a]Incubation mixtures contained 0.5 M Tris (pH 9), 0.02 M $MgCl_2$, 0.4 M $NH_4Cl$, and substrate, in a total volume of 3 mL. Reaction was initiated by addition of enzyme solution (20 μL), which was preassayed. Reactions were carried out at 30 ± 0.1 °C.
[b]Specific activity is 250 units $mg^{-1}$ (250 μmol $min^{-1}$ $mg^{-1}$ protein) under the assay conditions.[1] Error is ± 10% for all $V_{max}$ values.

The kinetic parameters for enzymic deamination were determined for (2S)-aspartic acid, (2S,3S)-3-methylaspartic acid and (2S,3S)-3-ethylaspartic acid by measuring the increase in uv absorbance at 240 nm. The results are shown in table 2.[11]

**Table 2**  Kinetic Parameters for the Deamination of 3-Substituted Aspartic Acids[a]

| Substrate | $K_m$ (mM) | [b]$k_{cat}(s^{-1})$ |
|---|---|---|
| 3-Methylaspartic acid | 2.37 ±0.2 | 822 |
| Aspartic acid | 10.5 ±0.82 | 5.8 |
| 3-Ethylaspartic acid | 17.08 ±1.4 | 365 |
| 3-Chloroaspartic acid | >50 | |

[a]Incubation mixtures contained at 0.5 M Tris (pH 9), 0.02 M $MgCl_2$, 0.001 M KCl, and substrate in a total volume of 3 mL. Reaction was initiated by addition of enzyme solution (20 μl). Reactions were carried out at 30 ± 0.1 °C.
[b]Error is ± 10% for all $V_{max}$ values.

The kinetic parameters for the amination reaction indicate that substrates possessing substituents smaller than the size of a propyl group react rapidly. Furthermore, the values of $V_{max}$ are similar. However, for the deamination reaction aspartate is a particularly slow substrate and there appears to be a step in the deamination reaction direction which is particularly sensitive to the size of the C-3 substituent. Since the observed rate is 137 times slower than for the physiological substrate and since a chemical step is thought (but not proven) to be rate limiting for this fast substrate, it is unlikely that a step other than a chemical step is rate limiting for the deamination of aspartic acid. Also, it should be noted that in the reverse reaction direction fumaric acid has a larger value for $V_{max}$ than the methyl homologue.

Two processes are involved in an $El_{cb}$-type elimination: removal of a proton from C-3 and then C-N bond cleavage, resulting in the expulsion of ammonia. In principle, either or both of these two processes would be expected to be sensitive to changes in the protein-substrate binding interaction adjacent to C-3. For example, a conformation change could result in a lower rate of proton abstraction from the C-3 position if the change decreased the acidity of the proton (through poor *aci*-carboxylate stabilization) or made the proton less accessible to the enzyme-bound base. Alternatively, or additionally, the C-N bond cleavage step could become more rate limiting through non-optimal alignment of the intermediate carbanion with the leaving N-atom (Scheme 3).

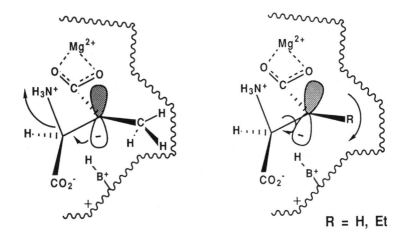

R = H, Et

**Scheme 3**      Non-optimal orbital alignment for aspartic acid

### 3   C-3 DEUTERIUM ISOTOPE EFFECTS

In order to gain some insight into which of these effects are important, the deamination reaction was studied using the C-3 deuteriated isotopomers of (2S)-aspartic, (2S,3S)-3-methylaspartic and (2S,3S)-3-ethylaspartic acid. These were prepared by conducting incubations similar to those described above in deuterium oxide.[9] Kinetic data was obtained using the methods described above at pH 9 (see Table 2). Comparison of the kinetic parameters revealed an isotope effect on V and V/K of 1.7 ± 0.3 for methylaspartic acid, an isotope effect of 1.16 ± 0.2 on V and V/K for ethylaspartic acid and no isotope effect for aspartic acid.[12,13] Thus the results suggested that C-3-H bond cleavage was not rate limiting for aspartic acid, implying that C-N bond cleavage is the slow step.

The finding that (2S,3S)-[3-$^2$H]-3-methyl-aspartic acid shows a significant primary isotope effect on V and V/K was particularly surprising, since it had been reported that an isotope effect could not be detected at pH 5.5, 7.5 or 10.5.[14] The conditions for these experiments, however, were not reported. Furthermore, the whole concept of the carbanion mechanism for enzymic deaminations had rested upon three major pieces of evidence, of which the lack of an observed isotope effect was one. Our findings, therefore, prompted us to critically evaluate the case for a carbanion mechanism.

### 4   THE CASE FOR A CARBANION MECHANISM

Three pieces of evidence collectively support the carbanion mechanism.

i.  The lack of an observed C-3 deuterium isotope effect.[14]

ii. The observation that substrate C-3 hydrogen exchange with the solvent could occur more rapidly than the deamination reaction.[14,15]

iii. The        observation that  the  enzyme  could
deaminate    the   L-*erythro*-   diastereomer,
(2S,3R)-3-methylaspartic   acid,   at  about  1%
of the  rate  of  the  L-*threo*-  diastereomer.[1]


     Since  we  were  able  to  show  that  an isotope
effect of 1.7 on V and V/K is observed over the pH
range  6.5-10.0,  we  were  confident  that  the  first
piece   of   evidence   supporting   the   carbanion
mechanism  could  be  discounted.    Thus,  we  turned
our  attention  to  the  exchange  reaction.

     In  order  to  assess  the  rate  of  exchange  of
H-3  of  substrates  with  the  solvent,  incubations  of
protio-  and  deuterio-  substrates  were  performed  in
tritium  oxide.  At  several  minute  intervals  up  to
1-2%  deamination,  as  judged  by  monitoring  $\Delta$OD  at
240  nm,  aliquots  of  the  incubation  mixture  were
removed,  quenched  in  acid  and  repeatedly  diluted
into  a  vast  excess  of  water,  and  then  lyophilised.
The  residual  radioactivity  from  the  samples  was
determined  by   scintillation  counting.

     Using  this  methodology,  it  was  possible  to
determine  that  the  rate  of  C-3  hydrogen  exchange,
$v_{ex}$ ,was  equal  to  $v_{deam}$  for  methylaspartic  acid  at
pH  9.0.  For  aspartic  acid  $v_{ex}$  was  equal  to  16
times  $v_{deam}$    at  pH  9.0.  At  pH  7.6  $v_{ex}/v_{deam}$  for
methylaspartic  acid   was  3.  Thus  there  was
evidence  to  confirm  that  C-3  hydrogen  exchange
could  occur  more  rapidly  than  the  deamination
reaction  for  both  methylaspartic  acid  and  aspartic
acid.  However,  when  the  ratio  $v_{ex}/v_{deam}$  was
determined  for  the  deuteriated  isotopomers  of  each
substrate  at  pH  7.6  and  9.0,  they  were  almost
identical  to  those  for  the  unlabelled  compounds,
*ie.*  the  isotope  effects  for  exchange  and
deamination  were  essentially  the  same.  These
results  are  extremely  difficult  to  rationalise  in
terms  of  tritium  wash-in  at  the  carbanion  level
and  indeed,  suggest  that  partitioning  between
exchange  and  deamination  occurs  after  C-N  bond
cleavage;  through  reverse  steps.

     In  spite  of  the  conceptual  difficulty  in
rationalising  the  above  results  in  terms  of  a

carbanion mechanism, the results of a rather elegant experiment by Bright still needed to be explained. Bright had shown that the ratio $v_{ex}/v_{deam}$ increased if ammonia was added to the incubation mixtures during deamination, and in further experiments showed that $^{15}$N- label from endogenously added $^{15}$N-ammonia did not become incorporated into the solvent hydrogen exchanged substrate pool.[14] Thus, from these experiments, it appeared that C-N bond cleavage did not occur prior to exchange. Nevertheless, two points need to be considered carefully. First, ammonium ion fulfils the role of potassium as a monovalent cationic activator, and, second, the order and rates of product desorption from the enzyme, if indeed there was an order, had not been determined.

Recently, we have determined that ammonia is an uncompetitive product inhibitor for the deamination of methylaspartic acid in the presence of saturating concentrations of potassium. At low potassium concentrations, ammonia acts as an activator, at low concentrations, and at high concentrations, as a mixed inhibitor. These results indicate that ammonia is the first product to be desorbed from the enzyme. Interestingly, methylamine is unable to fulfil either of the roles of ammonia and can neither act as an activator or a substrate (in the reverse direction). Indeed, it is a competitive inhibitor for amino acid substrates in the deamination direction and a competitive inhibitor for ammonia in the amination direction.

In further experiments, designed to probe the kinetics and order of product desorption in the deamination direction, $CD_3$-mesaconic acid and $^{15}$N-ammonia were incubated with unlabelled methylaspartic acid in the presence of the enzyme, see Scheme 4. Aliquots of the incubation reaction were removed at intervals of several minutes, and the label distribution of $^{15}$N and deuterium were measured in the substrate pool by mass spectrometry (multi-ion monitoring) after derivatisation, see Scheme 5. The ratios of $^{15}$N-ammonia and $CD_3$-mesaconic acid incorporated

into   the   substrate   pool   at   any   given   time,
reflected   the   ratios   at   the   start   of   the
experiment.      Thus,   it   is   not   possible   to   trap
either   of   the   products.      Since   it   is   known
independently   that   ammonia   desorbs   first,   the
results   indicate   that   the   rate   of   mesaconate
debinding   is   equal   to,   or   greater   than,   the   rate
of   ammonia   desorption.

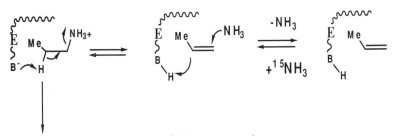

Compare N-isotope ratios in $CH_3$ compound

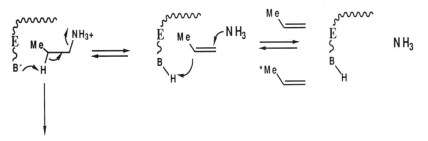

Compare Hydrogen isotope ratios in $^{14}N$ compound

**Scheme 4**   Experiments to trap products

## Isotope Distribution

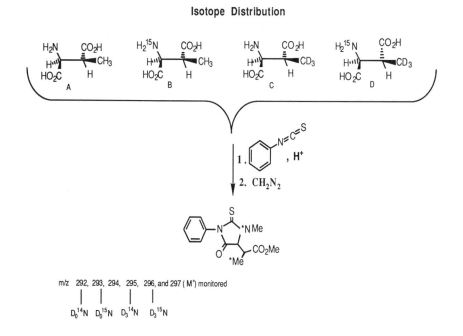

**Scheme 5**     Derivatisation of labelled methyl
                 -aspartic acids


          The third line of evidence in support of the
carbanion     mechanism,     the     observation     that
L-*erythro*- diastereomer of methylaspartic acid was
slowly deaminated by the enzyme,[1,2] has recently
been tested in our laboratory.  Although it has
been claimed that the *erythro*- and *threo*-
activities are inseparable, we have found that the
ability    of    the    enzyme    to    form
L-*erythro*-methylaspartic acid from mesaconic acid
and ammonia varies widely with fermentation batch.
Also, it should be noted that a biosynthetic
enzyme has been detected in *Acetobacter suboxydans*
which exclusively forms the *erythro*-diastereomer.
If, as indeed it appears, the *threo*-methylaspartase
is contaminated with small amounts of a different

enzyme  which  is  specific  for  *L-erythro-3-*
methylaspartate,  the  results  of  the
stereochemical  studies  would  not  support  the
carbanion  mechanism.


## 5    DOUBLE ISOTOPE FRACTIONATION

Since we were not compelled to believe that a
carbanion mechanism operates when methylaspartase
deaminates  its  physiological  substrate  on  the
basis  of  any  of  the  three  major  arguments  put
forward  by  previous  workers,  the  mechanism  was
investigated  further.  First,  the  importance  of
C-N bond cleavage was assessed by measuring the
$^{15}N$  isotope  effect  on  V/K  by  the  competitive
method.[17]

Accordingly,  natural  abundance  $^{15}N$-methylaspartic
acid  and  aspartic  acid  were  each  incubated  with
the  enzyme  until  10-20%  of  deamination  had
occurred.  The  ratio  of  nitrogen  isotopes  present
in  the  substrate  before  the  reaction  and  in  the
ammonia  produced  during  the  reaction  were
determined  by  isotope  ratio  mass  spectrometry
after  Kjeldahl  digestion  as  outlined  in  detail  in
scheme 6.

The  $^{15}(V/K)$  isotope  effect  for  methylaspartic
acid was  1.0246  ±0.0013  whereas  the  effect  for
aspartic  acid  was  1.0405  ±0.0015. Intrinsic  $^{15}N$-
isotope  effects  for  the  elimination  of  ammonia  are
expected  to  be  1.04-1.07,  since  $^{15}K_{eq}$  is  ~1.045
and  ~1.035[18]  for  the  enzymic  substrates
methylaspartic  and  aspartic  acid  respectively,  and
ammonia;  and,  since  the  isotope  effect  for  C-N
bond  formation  should  be  slightly  normal.[19]  Thus
it  appeared  that  C-N  bond  cleavage  was  at  least
partially  rate  limiting  for  aspartic  acid.  For
methylaspartic  acid,  however,  the  observed  effect
could  be  due  to  the  occurrence  of  a  slow  step
after  C-N  bond  cleavage,  for  example,  ammonia
de-binding,  which  would  bring  the  isotopically
sensitive  step  into  equilibrium.

Allow 10-20% reaction, then quench

Ammonia in the substrate is obtained by Kjeldahl digestion, mercuric sulphate oxidation followed by zinc dust reduction of the nitric acid produced.
The ammonia is distilled into sulphuric acid.

To determine the $^{15}N/^{14}N$ ratio, the ammonia is oxidized to dinitrogen using hypobromite The ratio is then measured using a VG SIRA 10 dual inlet isotope ratio mass spectrometer

$$^{15}(V/K) = \log (1-f) / \log( 1- [ fR/R_o])$$

where R and $R_o$ are the measured ratios ($^{15}N/^{14}N$) in the substrate and in the ammonia produced after the fraction of the reaction ( f ) respectively.

**Scheme 6** Determination of $^{15}(V/K)$ Isotope Effects

Cleland has devised a scheme to aid the interpretation of V/K isotope effects.[20] Where an isotopically sensitive step is flanked by a preceding slow step (*ie* .where the rate of the conversion of the intermediate back to free substrate is slower than the isotopically sensitive step) the observed value of the isotope effect on V/K is supressed towards unity. Thus, the step has a high *forward* commitment to reaction and little isotope fractionation is possible. Where the breakdown of the complex to give free product is slow compared to the reverse isotopically sensitive step, the step has a high *reverse* commitment and the isotope effect on V/K is suppressed (or enhanced) towards the equilibrium isotope effect.

Elegant work by the groups of Cleland[21] and Knowles[22] has allowed a simple test for

concertedness to be applied to enzymes which show
V/K isotope effects for two bond breaking
processes, one of which involves a hydrogen atom,
and one of which involves a heavy atom *eg.* C or N.
Essentially, the isotope fractionation for the
heavy atom is measured, as an isotope effect on
V/K, using protio substrate . The experiment is
then repeated using deuteriated substrate.If the
extent of heavy atom fractionation is reduced in
the presence of deuterium, then the forward or
reverse commitment to the heavy-atom bond cleavage
step must have increased. Thus, the reaction is
step-wise and it is possible to determine which
bond-breaking step occurs first.[20] If, however,
the reaction is concerted so that both
bond-breaking processes occur in the same step,
then the forward and reverse commitments remain
unchanged or decrease for the deuteriated
substrate and the observed value of V/K is
unaltered or increased. Since methylaspartase is
expected to operate by a balanced step-wise
carbanion mechanism, or by a concerted mechanism,
the change in fractionation of the the nitrogen
isotopes on the introduction of deuterium at C-3
can be anticipated for each mechanism, Scheme 7.

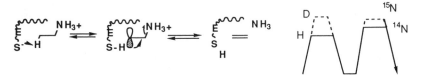

STEP-WISE

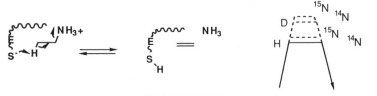

CONCERTED

**Scheme 7**            Double Isotope Fractionation for
                        Carbanion and Concerted reaction Pathways

When the deuteriated isotopomer of (2S,3S)-3-methylaspartic acid was incubated with the enzyme at pH 9.0, $^{15}(V/K)_D$ was 1.0241 ±0.0009 and therefore, identical to the value obtained for the protio substrate. These results indicate that the compound is deaminated *via* a concerted mechanism. When the double isotope fractionation experiment was performed on aspartic acid at pH 9.0, $^{15}(V/K)_D$ was 1.0278 ±0.0015 , lower than the value obtained for the protio substrate, indicating that the compound is deaminated *via* a step-wise process. Hence, the mechanism of deamination appears to change from concerted to step-wise when the methyl group at C-3 of the substrate is removed, *vide supra*.

There are, however, complications in the interpretation of the isotope fractionation data for both substrates.

First, since the $^{15}$N-isotope effect for protio methylaspartate is clearly not intrinsic, why does it not increase when the kinetic importance of the step is enhanced through the introduction of deuterium?
Second,how can a substrate (aspartic acid) which shows no primary deuterium isotope effect on V or V/K for deamination, or, indeed for C-3 hydrogen exchange, show a large isotope effect on nitrogen fractionation?

Both of these dichotomies can be explained by recalling that ammonium ion is an efficient surrogate for $K^+$ as a monovalent activator, and, furthermore, is a *product* in the forward reaction. In all of our studies in the forward deamination direction we have used 1 mM $K^+$, a subsaturating concentration ( note: at 50 mM $K^+$ at pH 9.0, $k_{cat}$ for methylaspartic acid is 5-fold higher than at 1 mM, V/K is 15-fold higher, and there is no detectable C-3 primary deuterium isotope effect on V or V/K). While these conditions are suitable for initial rate measurements, the enzymic reaction becomes autocatalytic if the products are allowed to accumulate. In the determination of $^{15}(V/K)$ isotope effects, up to 20% of reaction is allowed to occur. Thus, the forward and reverse commitments to the chemical step(s) must increase as their rates increase, as the reaction proceeds.

This analysis implies that the magnitude of $^{15}(V/K)$ isotope effects will be underestimated.

For methylaspartic acid then, the identity of $^{15}(V/K)$ and $^{15}(V/K)_D$ at pH 9.0 indicate that the chemical step is both *concerted* and *cleanly rate limiting*. Furthermore, this notion is supported by the identity of $^DV$ and $^D(V/K)$ over the pH range 6.5-10.0; by the observation that $^Dv_{ex}$ is equal to $^Dv_{deam}$ at any given pH although the actual ratio of $v_{ex}/v_{deam}$ varies; and, by the fact that at pH 6.5, where monovalent cations are less effective activators, the value of $^{15}(V/K)_D$, $1.0417 \pm 0.0010$ is ~1.6 times the value of $^{15}(V/K)$, $1.0255 \pm 0.0011$.

For aspartic acid, the value of $^{15}(V/K)$ was 1.040. Given that this value is an underestimate for the reasons outlined above, its true value must be very large indeed, and C-N bond cleavage must be cleanly rate limiting. Since the effect of ammonium ion is similar to that for methylaspartic acid, and since C-N bond cleavage is rate limiting, the effect of the monovalent cation must be to accelerate C-N bond cleavage. The observed decrease of $^{15}(V/K)$ for the deuteriated substrate indicates that the forward commitment to C-N bond cleavage has increased. This is entirely consistent with a mechanism whereby the monovalent cation enhances the rate of C-N bond cleavage to the point where it *approaches* the rate of C-H bond cleavage, *i.e.* , where the mechanism changes towards a concerted one. These ideas are currently being tested in our laboratory.

## 6   PROTEIN SEQUENCE

β-methylaspartase was reported to possess an (AB)$_2$ structure.[5,6] We have been unable to detect two different polypeptides and obtain a single polypeptide of 50,000 kD on denaturation with SDS-thiols, urea, or guanadine hydrochloride. We were able to sequence the N-terminal region of the protein without any complications and have prepared a 78-mer oligonucleotide probe. The probe has been used to screen libraries of

*Clostridium tetanomorphum* genomic DNA. A 1.5 kb Sau 3A fragment which encodes for the first 80 amino acids of methylaspartase has been sequenced, Figure 1,[23] and shows no homology with the superficially similar enzyme L-aspartase or, fumarase. Note that L-aspartase and fumarase show a considerable degree of protein sequence homology.[24]

Studies towards locating the gene for L-*erythro*-β-methylaspartic acid specific β-methylaspartase are also under way in our laboratory. Since the deamination product is also mesaconic acid, the reaction catalysed is a *syn*-elimination.

```
                                        •
GGACAGGTGAATAATTATGAAAATTGTTGACGTACTTTGTACACCAGGATTAACTGG
             M  K  I  V  D  V  L  C  T  P  G  L  T  G
                                                    14

ATTCTATTTTGATGACCAAAGAGCAATCAAAAAGGGAGCAGGACATGATGGATTTAC
 F  Y  F  D  D  Q  R  A  I  K  K  G  A  G  H  D  G  F  T
                                                    33

ATATACTGGCTCTACTGTAACAGAAGGATTTACTCAAGTAAGACAAAAAGGTGAATC
 Y  T  G  S  T  V  T  E  G  F  T  Q  V  R  Q  K  G  E  S
                                                    52

AATATCTGTATTATTAGTTCTTGCAAGATGGTCAN
 I  S  V  L  L  V  L  A  R  W  S  X
                                  64
```

**Figure 1**    N-Terminal Sequence of Methylaspartase[23]

7    SUMMARY

β-methylaspartase from *Clostridium tetanomorphum* has been used in the enantiospecific syntheses of 3-halogeno- and 3-alkyl- aspartic acids. The mechanism of the reaction catalysed by methylaspartase has been probed using an array of different substrates and isotope-dependent techniques. The results of these investigations clearly indicate that the physiological substrate is deaminated *via* a concerted mechanism, and that, under similar conditions, aspartic acid is deaminated *via* a stepwise mechanism. Thus, it appears that the methyl group of the physiological substrate ensures that the C-N bond is optimally aligned for cleavage in the transition state, and that removing the methyl group disturbs this alignment. To our knowledge, this is the first example of a concerted enzymic elimination process of this type. Since methylaspartase has hitherto been considered to be the archetypal example of enzymes which operate by a carbanion mechanism, this work calls for a re-evaluation of the mechanism of similar systems.

## ACKNOWLEDGEMENTS

The authors wish to thank Professor C. Greenwood and Mr. A. Thompson, Department of Biology, University of East Anglia, for growing *Clostridium tetanomorphum* H1; Professor A.A. Jackson, Department of Nutrition, University of Southampton, for performing isotope ratio mass spectrometry; Professor A. Atkinson and Dr. N. Minton, Centre of Applied Microbiological Research, Porton, for providing facilities suitable for cloning clostridial genes; and, M. Akhtar, M.A. Cohen, N. Thomas and S. Goda for their in-house contribution at Southampton. We thank the SERC for studentships to MA, MAC and NT, and for fellowships to SG and NPB, and the Royal Society for equipment grants and a University Fellowship.

## References

1. H. A. Barker, R. D. Smith, R. Marilyn, and H.Weissbach, *J. Biol. Chem.,*1959, **234**, 320.
2. V.R. Williams and J.G. Traynham, *Federation Proc.,* 1962, **21**, 247.
3. S. Uedao, K. Sato and S. Shimizu, *J. Nutri. Sci. Vitaminol.,* 1982, **28**, 21.
4. M. F. Winkler and V. R. Williams, *Biochim. Biophys. Acta.,* 1967, **146**, 287.

5. M. W. Hsiang and H. J. Bright, *J. Biol. Chem.*, 1967, **242**, 3079.
6. W. T. Wu and V. R. Williams, *J. Biol. Chem.*, 1968, **243**, 5644.
7. H.A. Barker, R.D. Smyth, E.J. Wawszkiewicz, M.N. Lee and R.M. Wilson, *Arch. Biochem. Biophys.*, 1958, **78**, 468.
8. M. Akhtar, M. A. Cohen, and D. Gani, *J. Chem. Soc. Chem. Comm.*, 1986, 1290.
9. M. Akhtar, N. P. Botting, M. A. Cohen, and D. Gani, *Tetrahedron*, 1987, **43**, 5899.
10. M. Akhtar, M.A. Cohen and D. Gani, *Tet. Lett.*, 1987, 2413.
11. N. P. Botting, M. Akhtar, M. A. Cohen, and D. Gani, *Biochemistry*, 1988, **27**, 2953.
12. N. P. Botting, M. Akhtar, M. A. Cohen, and D. Gani, *J. Chem. Soc. Chem. Comm*.,1987, 1371.
13. N. P. Botting, M.A. Cohen, M. Akhtar, and D. Gani, *Biochemistry*, 1988, **27**, 2955.
14. H. J. Bright, *J. Biol. Chem.*, 1964, **239**, 2307.
15. H. J Bright, R.E. Lundin and L.L. Ingraham, *Biochim. Biophys. Acta*, 1964, **81**, 576.
16. M.E. Maragoudakis, Y. Sekizawa, T.E. King and V.H. Cheldelin, *Biochemistry*, 1966, **5**, 2646.
17. M.H. O'Leary, 'Transition States of Biochemical Processes' ( Eds. R.D. Gandour and R.L. Schowen), 1978, p. 292, Plenum Press, New York.
18. J.D. Hermes, P.M. Weiss and W.W. Cleland, *Biochemistry*, 1985, **24**, 2959.
19. P.M.Weiss, P. F. Cook, J.D.Hermes, and W.W.Cleland, *Biochemistry*, 1987, **26**, 7378.
20. J.D. Hermes, C.A. Roeske, M.H. O'Leary and W.W. Cleland, *Biochemistry*, 1982, **21**, 5106.
21. J.D. Hermes, P.A. Tipton, M.A. Fisher, M.H. O'Leary, J.P. Morrison and W.W. Cleland, *Biochemistry*, 1984, **23**, 6263.
22. S.J. O'Keefe and J.R. Knowles, *Biochemistry*, 1986, **25**, 6077.
23. S. Goda, N.Minton, and D.Gani, unpublished results
24. S.A. Woods, J.S. Miles, R.E. Roberts and J.R. Guest, *Biochem. J.*, 1986, **237**, 5168.

# Structure and Dynamics of Ligand Binding to Enzyme Receptors

By D. J. Osguthorpe, P. Dauber-Osguthorpe, R. B. Sessions, and
P. K. C. Paul

MOLECULAR GRAPHICS UNIT, UNIVERSITY OF BATH, BATH, UK

## Introduction

The binding of ligands to receptors is an extremely important molecular recognition issue that has major implications for the design of drugs and insecticides. Unfortunately, most receptors are membrane bound proteins whose structure is unknown at present. However, certain drugs act on enzymes in the cytosol which are easier to crystallise and such systems provide models to study the recognition process with full structural information available.

The goal of molecular recognition studies is to understand how two molecules, such as a ligand and a receptor, interact in a specific manner. Although experimental studies can shed much light on the average structure, thermodynamics and kinetics of the process, in the final analysis they cannot give information on the forces involved, particularly for systems involving many different atoms in different chemical environments. Molecular mechanics calculations can give us a clue to the actual forces involved in a particular system, especially in conjunction with some experimental data for the system.

## Potential Parameters

All molecular mechanics computations are based on representing the potential energy surface of the system using a functional form such as:

$$V = \Sigma\{D_b[1 - e^{-\alpha(b-b_0)}]^2 - D_b\} + 1/2 \Sigma H_\theta(\theta - \theta_0)^2 \qquad (1)$$

$$+ 1/2 \Sigma H_\phi(1 + s \cos n\phi) + 1/2 \Sigma H_\chi\chi^2$$

$$+ \Sigma\Sigma F_{bb'}(b - b_0)(b' - b_0')$$

$$+ \Sigma\Sigma F_{\theta\theta'}(\theta - \theta_0)(\theta' - \theta_0') + \Sigma\Sigma F_{b\theta}(b - b_0)(\theta - \theta_0)$$

$$+ \Sigma F_{\phi\theta\theta'} \cos \phi(\theta - \theta_0)(\theta' - \theta_0') + \Sigma\Sigma F_{\chi\chi'}\chi\chi'$$

$$+ \Sigma\epsilon[(r^*/r)^{12} - 2(r^*/r)^6] + \Sigma q_i q_j/r$$

This functional form is based on the Valence Force field used to analyse vibrational spectra[1] and is the functional form used at Bath. It reflects the energy necessary to stretch bonds (b), distort angles ($\theta$) from their unstrained geometries, and twist torsion angles ($\phi$). In addition, as is known from vibrational spectroscopy and normal mode analysis, these internal deformations are coupled, and this is represented by the cross terms (terms containing two internals, e.g. b and b', or b and $\theta$). Finally, the non bonded (or Lennard-Jones) and coulomb interactions, representing steric repulsions, dispersion or attractive forces, and electrostatic interactions are given by the last three terms. Similar force fields are the basis of all current programs that perform molecular mechanics calculations, such as MM2,[2] CHARMM,[3] AMBER,[4] GROMOS,[5] and DISCOVER.[6]

There are two aspects of the potential used which determine its ability to reproduce experimental data, the functional form and the potential parameters. The functional form determines the shape of the energy surface and a given functional form may not be able to reproduce all deformations correctly. For example, the Valence Force field we use contains cross-terms. These have been shown to be necessary to reproduce vibrational spectra and are important in order to be able to reproduce the dynamics of the system. Recently, we have added a bond-torsion cross-term to the above force field in order to be able to reproduce the abnormal bond lengths in anomeric systems.[7] The potentials used in the programs mentioned above differ somewhat in the exact terms used in the potential function.

The second aspect is the values of the potential parameters, which is where the above force fields differ most. In general, the potential parameters are determined by varying the potential parameters until calculated properties reproduce experimental data. This procedure was developed by Lifson[8] to produce a Consistent Force field in which the parameters are fitted using a least square procedure such that they are the optimum parameters for the molecules fitted in the least square sense.

The parameters we are using for the Valence Force Field were determined by fitting a wide range of experimental data including crystal structures (unit cell vectors and orientation of the asymmetric unit), sublimation energies, molecular dipole moments, vibrational spectra and strain energies of small organic compounds. *Ab initio* molecular orbital calculations have also been used in conjunction with experimental data to give information on charge distributions (used to derive partial atomic charges), energy barriers and coupling terms.[9-12]

The accuracy of any calculation with such a force field depends on how well the functional form and potential parameters represent the actual energy surface of the molecule. Moreover, depending on the functional form chosen and the experimental data used in fitting the potential parameters, a given potential will reproduce some physical properties well and others poorly. For example, most available experimental data is for molecules that are not strained and so a set of potential parameters derived from such data cannot be expected to reproduce highly strained conformations with the same degree of accuracy.

*Properties Available from the Energy Surface*

The molecular energy surface as described by equation (1) implicitly contains all information concerning the conformational and statistical mechanical properties of the biomolecular system. The local conformational minima of the system can be determined by energy minimisation and thus the relative energy differences of the minima. The force on each atom is given by the negative of the first derivative of the potential energy with respect to cartesian coordinates. With this information, and given a starting structure and velocities, Newton's equations of motion can be numerically solved for the system to generate a trajectory of the dynamics of the system under the forces on the atoms. This trajectory contains all information both on the dynamic behaviour of the molecule and its thermodynamic behaviour (by computing statistical mechanical averages over the trajectory) in the classical limit. We can obtain statistical thermodynamic, spectral and structural properties as appropriate time averaged quantities, while dynamic properties such as structural fluctuations or conformational transitions may be monitored directly by analysis of the trajectory or by viewing the conformational motion on an interactive graphics system or from a movie of the same.

*Structure and Dynamics of Ligand Binding to Receptors*

As mentioned in the introduction, for many systems of interest only the ligand structure is known, either in the crystal form via X-ray crystallography or in solution using n.m.r. techniques such as NOE, none if which necessarily represent the structure of the ligand when bound to the receptor. In such cases it may be possible to gain some insight into the binding conformation of the ligands by searching for plausible accessible conformations of ligands whose biological activity is known, both agonists and antagonists.[13]

*Accessible Conformations of MCH and MCH Fragments.*

Melanin concentrating hormone (MCH) is a neuropeptide produced in the hypothalamus. In teleosts it concentrates melanin within the pigment cells of the skin,[14] hence causing the fish skin to appear paler, but it is not a simple antagonist of melanin stimulating hormone (MSH). It also induces melanosome dispersion within tetrapod melanophores.[15] MCH also acts as a potent pituitary hormone, inhibiting the release of ACTH in mammals,[16] and stimulating growth hormone release in rats.[17] Although it has been shown to be present in man its function in man is, as of yet, unknown.

MCH is an oligopeptide of 17 residues with the sequence:

Asp-Thr-Met-Arg-Cys-Met-Val-Gly-Arg-Val-Tyr-Arg-Pro-Cys-Trp-Glu-Val

A disulphide bridge between $Cys^5$ and $Cys^{14}$ forms an intramolecular ring of 10 residues.

As the first stage in determining the shape and electrostatic requirements of the MCH receptor, we have performed theoretical analyses of the conformational features of the intact peptide, the cyclic decapeptide, $MCH_{5-14}$, and the linear, reduced decapeptide, using molecular dynamics simulations followed by energy minimisation.[18, 19] In parallel, the two peptides plus a range of fragments of the peptide have been synthesised by the Organic Chemistry group at Bath, for use in biological testing and for n.m.r. conformational analysis.[18, 20]

We have performed a total of 150 picoseconds of dynamics so far on MCH and the 2 fragments, the cyclic $MCH_{5-14}$ ring and the linear $MCH_{5-14}$ fragment. In all cases under consideration we found that the largest conformational changes occurred in the $Gly^8$-$Arg^9$ region. This shows a frequent conformational transition of the Gly residue from $\alpha$-helical to $\gamma$-turn. This in turn affects the conformation of its immediate neighbour Arg, and the conformational changes of $\psi_{Gly}$ and $\phi_{Arg}$ occur in tandem. In terms of the conformational parameters and backbone structural features the $(\phi,\psi)$ of Gly and Arg move from around (-80,40) and (--150,100) to (-70,-40) and (-80,100) respectively. Thus there is a competition between the formation of a $\gamma$ turn at Gly and a type II $\beta$ turn across the adjacent $Arg^9$-$Val^{10}$ junction. Surprisingly the two Val residues on either side of the Gly and Arg residues do not show any conformational changes.

Some regions of the peptide were seen to be constrained in the simulations. One such area was the section $Val^{10}$ to $Cys^{14}$, around the $Pro^{13}$ residue. However, unusually the Pro is not in a $\beta$-turn conformation which one normally expects in peptides.[21] Instead, the Pro is in a polyproline conformation.

The final major feature seen in the dynamics, and in the accessible local minima generated from the trajectory, is an internal cross-ring hydrogen bond from the Tyr[11] side chain hydroxyl to the backbone carbonyl of Cys[5], see figure 1.

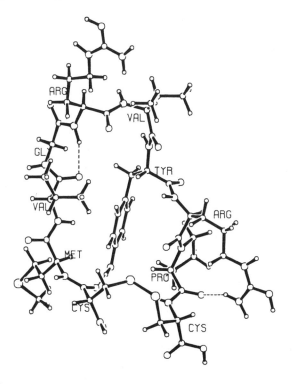

Figure 1. Sample minimised conformation of cyclic $MCH_{5-14}$ showing the cross-ring hydrogen bond.

Thus, we have defined constrained features and regions of conformational flexibility of MCH by analysis of molecular dynamics trajectories. Recently these conformational features have been shown to exist in independent experimental (see references 18, 19 and 20 for full details).

At this stage of the investigation, we cannot determine which of the conformations may be the binding or active conformations. Additional information, from a known antagonist or additional agonists or inactive compounds, is needed before we can make a more specific prediction of the active conformation. However, the conformational features determined by molecular dynamics are being used to direct further synthetic studies which are designed to lock in one or the other of the features. For example, replacing the Gly with a D-amino acid residue, will constrain the flexibility of the peptide in this region, and enhance the preference of the $\alpha$-helical over the $C_7$ equatorial conformation. This emphasises the close relationship that is necessary between the synthetic, biological and the theoretical studies to arrive at putative binding or active conformations when the structure of the receptor is unavailable.

## Structure of Ligand Binding to Enzyme Receptors

We have been studying phospholipase $A_2$ as a system where the structure of a model of the receptor is known. Intracellular membrane phospholipases $A_2$ has a primary role in the inflammatory response by liberating arachidonic acid, from membrane phospholipids, which is further metabolised to prostaglandins prostacyclin, and leukotrienes. Although very little is known about intracellular phospholipase, there are indications that they are closely related to the secretory enzymes.[22] X-ray structures for the secretory enzymes are known and so these can provide the basis for investigating ligand binding and the recognition problems associated with phospholipids in a ligand-receptor system where the structure of the receptor is known.

We studied the binding of substrate, the corresponding tetrahedral intermediate and of some known and new inhibitors. The modelling of the complexes was based on the X-ray structure of the pancreatic apoenzyme $PLA_2$.[23] Inspection of this structure as well as the other known $PLA_2$ structures, reveals a cleft close to the Ca binding loop, and containing the hydrogen bonded residues His[48] and Asp[99], reminiscent of the catalytic centre in serine proteases. There is also experimental evidence that the active site is in this cleft region.[24] Thus we started our modelling by docking ligands into this cleft, in a manner consistent with the proposed catalytic mechanism.

Before the docking could begin we first refined the structure of the enzyme itself. This was done by minimising the energy of the enzyme with respect to *all* degrees of freedom using a force field developed for peptides and proteins.[12] All hydrogen atoms were built on to the heavy atoms. X-ray waters in the active site cavity as well as around charged residues were included explicitly. In addition all water molecules which are bridging two residues were also included since they are important for maintaining the structural integrity of the protein. Additional water molecules were generated to fill up the active site (10Å from His[48]) and around all charged residues (3.5Å radius). The system was minimised until the the derivatives of the energy with respect to the cartesian coordinates was less than 0.02 kcal/Å. In order to evaluate the success of our methods in simulating experiment, the minimised structure was compared to the X-ray structure. Figure 2 shows a superposition of the minimised and X-ray structures. As can be seen all the structural features were maintained. The root mean squares deviation (RMSD) between all heavy atoms in the two structure was 1.25Å. However most of the deviations were in the side chains, the RMSD of the main chain heavy atoms is 1.05Å. Furthermore, the RMSD of the secondary structure (helices and $\beta$ sheet) is only 0.8Å. A detailed examination of the RMSD as a function of residue, and comparison to the experimental temperature factors, reveals that most deviations are in surface residues with high temperature factors. In addition, it is to be expected that the absence of the crystalline environment in the simulation will lead to deviations from experiment for surface residues. The deviations of the secondary structure atoms is of the same size as deviations between experimental structures of similar proteins in different crystal environments.[12] These results indicate that the potentials and techniques simulate experiment adequately and give us confidence in applying them to additional systems and analysing the results in terms of the interatomic forces.

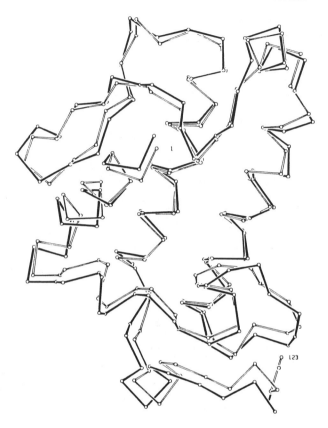

Figure 2. Superposition of the PLA$_2$ X-ray structure (open bonds) and the minimised structure (filled bonds) showing the C$\alpha$ atoms only.

The first ligand we docked into the cavity of the minimised enzyme was a phospholipid. Observed phospholipid conformations fall into two broad classes, those with the alkyl chains either parallel or perpendicular to the glycerol chain. Only one of these, the parallel conformer could be docked successfully. The final docked substrate structure is thus similar to an X-ray structure and some solution and membrane conformations.[25-27] In addition to selecting the appropriate substrate conformation we found it necessary to alter the torsion angles of some of the sidechains of the enzyme's residues. These included Leu[19], Leu[31] and Tyr[69] which in the experimental structure are oriented across the cleft, partially blocking the entrance to the active site. In addition the side chain of Asp[49] had to be moved away from the Ca ion to enable coordination of the lipid's phosphate group to the metal. The complex was minimised to relieve any new strain introduced in the docking and to optimise enzyme- ligand interactions. The resulting structure maintained the structural features of the apoenzyme and lead to favourable interactions between the ligand and enzyme. The phosphate and the acyl carbonyl are stabilised by interacting with the calcium ion. In addition the acyl group to be hydrolysed is positioned next to an X-ray water which hydrogen bonds to the His[48] residue. This arrangement is consistent with the suggested catalytic mechanism at the beginning of the reaction path. As mentioned above, the side chains of some residues had to be moved to enable docking of the substrate. Once the minimisation was complete, inspection of the resulting complex revealed that an even better interaction could be achieved by readjusting the positions of some of the residues at the cavity entrance. For example, we manually changed the torsions of Tyr[69] and reminimised, resulting in closure of the active site entrance and the hydroxyl group hydrogen bonding to the ester carbonyl of the phospholipid. The active site region of this final model of the enzyme-ligand complex is shown in figure 3. The necessity to change the conformation of parts of the protein highlights the problems with modeling protein-ligand interaction using a rigid "lock and key" concept. In particular in proteins with large hydrophobic cavities, such as phospholipase, structural rearrangement will occur in the absence of ligand to exclude water from the cavity. Thus it is necessary to treat both ligand and enzyme as flexible entities by allowing *all* degrees of freedom to change during a molecular mechanics minimisation, or in molecular dynamics simulations.

The next stage of modeling involved docking inhibitors into the active site, to test and extend the model of enzyme - ligand binding. This included known inhibitors (like anilinenaphthalene sulphonic acid, ANS),

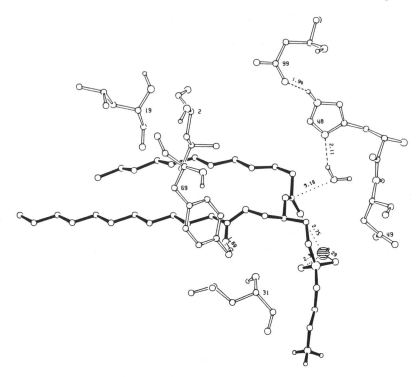

Figure 3.    The minimised complex of PLA$_2$ (open bonds) with the docked phospholipid (filled bonds) and the Ca$^{2+}$ ion (hatched). showing hydrogen bond distances in Å.

the inhibitors of Ripka[28] and Sigler.[29]

In parallel, a novel substrate analogue, 3-arachidonyl-4(*O*-phosphoethanolamino)-methyltetrahydrofuran-2-one, had been synthesised in the Organic Chemistry Group at Bath[30] and we docked this compound into the active site as well. Later, this molecule was shown to be a more potent inhibitor of PLA$_2$ than ANS. The molecule was synthesised as the racemate and the tetrahydrofuranone part of the molecule, could, in principle, adopt either *cis* or *trans* conformations or exist as *E*- or *Z*- exocyclic enols. Eight chiral isomers are thus possible.

Modelling studies were performed for each in order to determine which of them might be compatible with the ligand binding model derived above. The conformations of the novel analogue were adjusted to mimic the defined substrate requirements by constrained minimisations and compared with the model substrate in terms of spatial fit and favourable orientation of functional groups. The best fit to our substrate model was achieved with the 3S,4R isomer of the novel analogue. The two next best fits, E,4R and 3R,4R, both have the plane of the lactone ring perpendicular to that defined by the phospholipid side chains. The poorest fits are observed with isomers having 4S stereochemistry. We therefore predict that for optimal activity in structural variants based upon this novel analogue, 3S,4R- stereochemistry will be required.

In order to gain insight into the enzyme's mechanism of action, and as a preliminary stage for investigating inhibitors which are analogues of the tetrahedral intermediate, we studied the complex of the enzyme and the hydrolysis intermediate of the phospholipid substrate. While the charge separation model still resembles the the initial structure, with hydrogen bonding between the intermediate and the His (see figure 4). In the charge relay model, the two oxygens of the reaction centre moved further from the His and interacts with the Ca ion. These models have been used to examine the docking of the tetrahedral intermediate isosteres of Gelb.[31, 32]

## Dynamic Aspects of Ligand Binding to Enzyme Receptors

During the modelling of ligand binding to the enzyme, we became interested in the low-frequency motions of the enzyme, which could give rise to the types of structural changes that were made manually to the active site when docking the substrate. The most important motions of biological molecules are the low frequency collective motions associated with large fluctuations in conformation and conformational transitions. These motions lead to important biological phenomena such as domain closure upon substrate binding, allosteric effects, thermostability of enzymes, and may be important in the molecular recognition pathway of

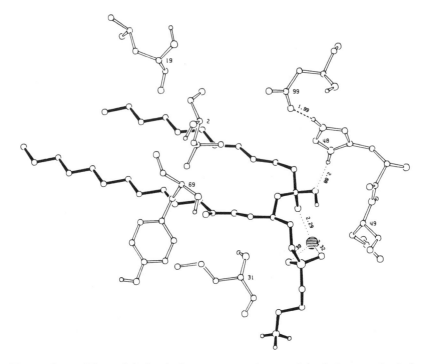

Figure 4.    The minimised charge separation model of the tetrahedral
intermediate of hydrolysis with hydrogen bonding distances
in Å (PLA$_2$ open bonds, phospholipid filled bonds and Ca$^{2+}$
ion hatched).

ligand docking.

Molecular dynamics simulations have become an important tool for
understanding the behaviour of biomolecules. Although the molecular
dynamics trajectory represents a comprehensive and realistic description
of the molecule, it is difficult to interpret this wealth of  information due
to the complexity of the motion.  In particular, it is difficult to focus on
the low frequency conformational motions in the molecular dynamics trajectory due to the superimposed high frequency local motions.  We have

developed a novel method for analysing molecular dynamics trajectories, which facilitates focusing on the low frequency motions of proteins.[33] By treating the trajectory as a "signal", we can apply digital signal processing techniques to the trajectory and, for example, "filter out" the high frequencies.

We have used the filtering technique to study the dynamics of $PLA_2$. We generated a molecular dynamics trajectory of the apoenzyme (including waters) at 300K, starting from the minimised structure. The trajectory of 2,705 atoms was filtered so that all frequencies higher than 50 $cm^{-1}$ were removed. The filtering process eliminated the high frequency bond and angle fluctuations and the remaining motion is mainly torsional, including conformational transitions. The analysis of the filtered trajectory revealed interesting characteristic motions of the protein, including concerted movements of helices, and changes in shape of the active site cavity. These features can be seen in a movie of the filtered trajectory we have made.

One of the aspects we were interested in was the motions around the active site cavity, where we had made significant changes in docking the substrate. The four residues around the entrance of the active site, $Tyr^{69}$, $Leu^{31}$, $Leu^{19}$ and $Leu^2$ exhibited high mobility with a standard deviation of $\approx 0.8$-$1.0$Å. In particular, the $Tyr^{69}$ and $Leu^{31}$ residues move in such a way as to open and close the entrance to the cavity. Surprisingly, most of this motion is due to backbone motion and not sidechain motion. This contrasts sharply with the procedures normally used in docking in which the sidechains are rotated in order to fit the new molecule in. The molecular dynamics simulation reveals that significant changes in shape can be obtained by coupled changes in backbone torsion angles.

The other aspect of the dynamics we investigated was the motion of the secondary structures, in particular the two anti-parallel helices which carry the catalytic $Asp^{99}$-$His^{48}$ couple and of the N-terminal helix, which is implicated in the enhanced activity of the enzyme on micellar substrates. The pair of anti-parallel helices are connected by three disulphide bridges and form a relatively rigid unit as seen from the filtered trajectory.

The two helices undergo extension and compression along their helix axes with very little relative motion between the helices. This mode of motion retains the relative orientation of the $Asp^{99}$ and $His^{48}$ catalytic residues.

The N-terminal helix is relatively flexible and exhibits a twisting motion. The first residue of this helix, $Ala^1$ plays a crucial role in the enhanced activity of the enzyme on membrane surfaces or micelles.[34] The $Ala^1$ is connected to the catalytic network via hydrogen bonds yet it fluctuates significantly more compared to other residues of the active site, for example the maximum deviation between this residue and $Asp^{99}$ is 1.0Å compared to a 0.4Å maximum deviation between the $His^{48}$-$Asp^{99}$.

*Summary*

Empirical potential energy calculations can give information on the interatomic forces involved in the molecular recognition of a ligand by a receptor. The validity of such calculations is crucially dependent on the potential parameters used in the calculations. Using energy minimisation techniques we have created a model for ligand binding to phospholipase $A_2$ which has been used to investigate the structure and energetics of the binding of known inhibitors and a novel analogue. A novel method for the analysis of conformational information in a molecular dynamics trajectory has been used to study how the dynamic motion of the system could affect the molecular recognition of the ligand in the phospholipase system. Molecular dynamics was also used to generate conformations of MCH to show how calculations can aid in investigating molecular recognition when the structure of the receptor is unknown.

*References*

1.  R. G. Snyder and J. H. Schachtschneider, *Spectrochim. Acta*, **19**, 85-116 (1963).

2.  U. Burkert and N.L. Allinger, in *Molecular Mechanics*, American Chemical Society, Washington, D.C. (1982).

3.  B. R. Brooks, R. E. Bruccoleri, B. D. Olafson, D. J. States, S. Swaminathan, and M. Karplus, *J. Comp. Chem.*, **4**, 187 (1983).

4.  S. J. Weiner, P. A. Kollman, D. A. Case, U. C. Singh, C. Ghio, G. Alagona, S. Profeta, Jr., and P. Weiner, *J. Am. Chem. Soc.*, **106**, 765-784 (1984).

5.  H.J.C. Berendsen, J.P.M. Postma, W.F. van Gunsteren, A. DiNola, and J.R. Haak, *J. Chem. Phys.*, **81**, 3684-3690 (1984).

6.  Available from Biosym Technologies, San Diego, USA

7.  R.C. Viner and D.J. Osguthorpe. to be published

8.  S. Lifson and A. Warshel, *J. Chem. Phys.*, **49**, 5116 (1968).

9.  A. T. Hagler, E. Huler, and S. Lifson, *J. Am. Chem. Soc.*, **96**, 5319 (1974).

10. S. Lifson, A. T. Hagler, and P. Dauber, *J. Am. Chem. Soc.*, **101**, 5111 (1979).

11. P. Dauber and A. T. Hagler, *Accts. of Chem. Res.*, **13**, 105 (1980).

12. P. Dauber-Osguthorpe, V. A. Roberts, D. J. Osguthorpe, J. Wolff, M. Genest, and A. T. Hagler, *Proteins: Structure, Function, and Genetics*, **4**, 31-47 (1988).

13. E.L. Baniak, L.M. Gierasch, A.T. Hagler, J. Rivier, T. Solmajer, and R.S. Struthers, in *Program of the Ninth American Peptide Symposium*, *Toronto, Canada*, p. 125 (1985).

14. I.D. Gilham and B.I. Baker, *J. Endocrin.*, **102**, 237 (1984).

15. B.W. Wilkes, V.J. Hruby, W.C. Sherbrooke, A.M. Castrucci, and M.E. Hadley, *Science*, **224**, 1111 (1984).

16. B.I. Baker, D.J. Bird, and J.C. Buckingham, *J. Endocrin.*, R5-R8 (1985).

17. G. Skotfitsch, M. Jacobowitz, and N. Zamir, *Brain Res. Bull.*, 635 (1985).

18. D.W. Brown, M.M. Campbell, R.G. Kinsman, C. Moss, P.K.C. Paul, D.J. Osguthorpe, and B. Baker, *J Chem. Soc. Chem. Commun.*, 1543-1545 (1988).

19. P.K.C. Paul, P. Dauber-Osguthorpe, M.M. Campbell, D.W. Brown, R.G. Kinsman, C. Moss, and D.J. Osguthorpe, *Biopolymers.* in press

20. B.I. Baker, D.W. Brown, M.M. Campbell, R.G. Kinsman, C. Moss, D.J. Osguthorpe, P.K.C. Paul, and P. White, *Biopolymers.* in press

21. G.D. Rose, L.M. Gierasch, and J.A. Smith, *Adv. Protein Chem.*, **37**, 1 (1985).

22. M. Okamoto, T. Ono, H. Tojo, and Y. Yamano, *Biochem. Biophys. Res. Commun.*, **128**, 788 (1985).

23. B.W. Dijkstra, K.H. Kalk, W.G.J. Hol, and J. Drenth, *J. Mol. Biol.*, **147**, 97-123 (1981).

24. H.M. Verheij, J.J. Volwerk, E.H.J.M. Jansen, W.C. Puyk, B.W. Dijkstra, J. Drenth, and G.H. de Haas, *Biochemistry*, **19**, 743-750 (1980).

25. M. Elder, P. Hitchcock, R. Mason, and G.G. Shipley, *Proc. R. Soc. Lond. A*, **354**, 157-170 (1977).

26. R.H. Pearson and I. Pascher, *Nature*, **281**, 499-501 (1979).

27. H. Hauser, W. Guyer, I. Pascher, P. Skrabal, and S. Sundell, *Biochemistry*, **19**, 366-373 (1980).

28. W.C. Ripka, W.J. Sipio, and J.M. Blaney, *Lectures in Heterocyclic Chem.*, **9**, S95-S109 (1987). Suppl. to J. Het. Chem 24

29. A. Achari, D. Scott, P. Barlow, J.C. Vidal, Z. Otwinowski, S. Brunie, and P.B. Sigler, *Cold Spring Harbor Symposia on Quantitative Biology*, **LII**, 441-452 (1987).

30. M.M. Campbell and M. Sainsbury. in preparation

31. M.H. Gelb, *J. Am. Chem. Soc.*, **108**, 3146-3147 (1986).

32. W. Yuan and M.H. Gelb, *J. Am. Chem. Soc.*, **110**, 2665-2666 (1988).

33. P. Dauber-Osguthorpe, R.B. Sessions, and D.J. Osguthorpe. submitted

34. A.J. Slotboom, M.C.E. van Dam-Mieras, and G.H. de Haas, *J. Biol. Chem.*, **252**, 2948-2951 (1977).

# Protein Engineering: Achievements and Prospects

By D. M. Blow

BLACKETT LABORATORY, IMPERIAL COLLEGE OF SCIENCE, TECHNOLOGY AND MEDICINE, LONDON SW7 2BZ, UK

## 1 INTRODUCTION

The aminoacyl-tRNA synthetases provide an ideal system for the study of enzyme specificity. Accurate protein synthesis relies on the faithful recognition of a specific amino-acid and a specific tRNA synthetase. The observed fidelity of protein synthesis requires that the sum of the error rates in these two processes, plus the error rates in nucleic acid transcription and in recognition processes during chain elongation should be less than 1 in 3000. In practice we can be sure that the fidelity of the recognition processes by the aminoacyl-tRNA synthetases must be at least $10^4$.

Regrettably, despite intense efforts by several groups of x-ray crystallographers, there are only two detailed three-dimensional structures available for amino-acyl tRNA synthetases : tyrosyl-tRNA synthetase[1,2] and methionyl-tRNA synthetase[3,4]. For tyrosyl-tRNA synthetase, there is an accurate crystallographic model for the binding of the tyrosine substrate[5,6], but no crystallographic evidence about the binding of tRNA exists. For methionyl-tRNA synthetase the methionine-binding site cannot be identified in the current structure, and it presumably means that a conformational change occurs when methionine is bound. The only ligand for which there is firm structural data about the binding is ATP[7].

Our best model for amino-acid recognition is therefore the crystal structure of tyrosyl-tRNA synthetase[2], and the slightly more precise structure of the truncated mutant of the enzyme[6], in which the amino-acid binding domain appears completely unaltered in structure, but a disordered domain of 100 amino-acids has been removed by protein engineering[9].

Some aminoacyl-tRNA synthetases, notably the valyl enzyme, achieve their fidelity of aminoacylation by an "editing" mechanism in which a second recognition step occurs, and incorrectly charged aminoacyl-tRNA's are hydrolysed[9]. There is no evidence that such a process occurs in tyrosyl-tRNA synthetase, and the selectivity of the enzyme for tyrosine against its most important competitor, phenylalanine, is probably achieved by a single recognition process[10].

This enzyme also presents an interesting example of a binding site for a substrate (ATP) which is bound weakly, but whose intense reactivity makes it appear to have a fairly respectable Michaelis constant $K_M$. A wide range of protein engineering experiments have been made on the enzyme, coupled with careful kinetic measurements which give which give important information about the energetics of the enzyme-substrate interactions[10-18]. A number of these mutants have also been studied by x-ray diffraction[19-22].

## 2 BINDING OF THE TYROSINE SUBSTRATE

A striking feature of the surface of the tyrosyl-tRNA synthetase molecule is a deep cleft which forms the outer part of the enzyme active site. From the bottom of this cleft there opens a narrow slot which forms the tyrosine binding site (Figure 1a). The bottom of the slot is so deeply buried within the enzyme that it lies virtually at the centroid of the substrate-binding domain. The slot is deep enough that the carboxylate group of a bound tyrosine molecule only just emerges from it; it is sufficiently narrow that both sides of the phenolic ring are in van der Waals contact with adjacent amino-acids (Figure 1b); but in the third dimension it is wider than is necessary to accommodate a tyrosine molecule (Figure 1c).

The phenolic hydroxyl group of the tyrosine substrate is bound at the very bottom of this slot, and lies just at the centre of the substrate-binding domain. It is in a highly polar environment, because two polar groups, the carboxylate of Asp176 and the hydroxyl of Tyr34, are favourably oriented to form hydrogen bonds with it. In the absence of substrate, they are bridged by a water molecule.

One side of the pocket is formed by Gly36, part of the
phenyl ring of Tyr34, the side-chain of Leu68, and the
alpha carbon of Gly70. This side of the pocket is
therefore relatively non-polar. If residues 36 and 70
were not glycine, their side-chains would protrude into
the pocket and prevent binding.

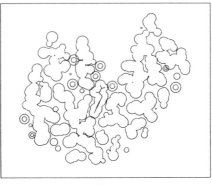

( a )

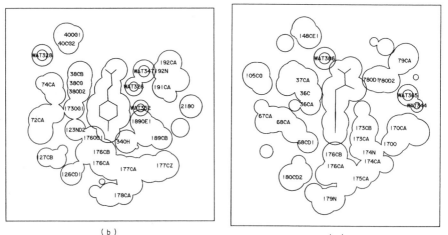

( b )                                              ( c )

Figure 1 Sections through a space-filling model of the
substrate-binding domain of tyrosyl-tRNA synthetase[6].
Tightly bound water molecules are shown as double circles.
a Whole domain.  b Section through the plane of the
aromatic ring of the substrate.  c Section perpendicular
to this plane.

The other side of the pocket is formed predominantly
of polar side chains, Thr73, Tyr169, Gln173, Asp176,
Gln189 and Gln195, which form an extensive hydrogen-bonded
network. At one edge of the pocket there is a free volume
adjacent to the tyrosine ring and close to Cys35, Ile190,
Gly191 and Gln195. When tyrosine is present, three
well-ordered water molecules are observed in this free
volume (Figure 1c).

It is at first sight surprising that a binding site
which has to give a high specificity towards tyrosine
should offer a relatively polar surface to the aromatic
ring, and that it should be so poorly shaped at one edge
that three water molecules are accommodated. The first
observation becomes less perplexing when it is realised
that *in vivo* the chief competitor for correct binding in a
tyrosine-binding pocket will be phenylalanine, whose
aromatic system is far less polarisable than that of
tyrosine. A polar environment for the aromatic ring will
favour tyrosine over phenylalanine. Moreover, a highly
non-polar surface would repel the tyrosine hydroxyl group
as it entered the slot, increasing the activation energy
of binding. It is also likely that the over-wide slot
provides a convenient route for the deeply buried water
molecules in the bottom of the slot to escape, as the
tyrosine advances towards the bottom. This effect will
also reduce the activation energy of binding.

## 3 PROTEIN ENGINEERING THAT AFFECTS TYROSINE BINDING

It appeared likely that the replacement of Asp176 by
a non-polar amino-acid, especially if coupled by the
alteration of Tyr34 to phenylalanine, would remove the
highly polar environment for the hydroxyl of the substrate
tyrosine, and enhance the possibility of binding
phenylalanine in its place. This proved not to be
practicable due to difficulty of expression of the enzyme
whenever Asp176 was altered.

A mutant in which Tyr34 was replaced by phenylalanine[10], led to an increased dissociation constant for
tyrosine, corresponding to an interaction energy of 0.52
kcal/mol [16]. As predicted, this mutation had a
considerable effect on the power of the enzyme to
discriminate against phenylalanine, reducing the relative
specificity by a factor 15.

The alpha-amino group of the substrate forms hydrogen
bonds to three polar side-chains, Asp78, Tyr169 and

Gln173, so orientated as to form a tetrahedral coordination around the nitrogen. When Tyr169 was changed to
phenylalanine, a considerably larger effect was observed
on tyrosine binding, corresponding to the enhanced energy
of a hydrogen bond involving a charged group. The change
of interaction energy was 2.58 kcal/mol [16].

## 4 BINDING OF THE ADENYLATE MOIETY

The first step of the reaction catalyzed by tyrosyl-
tRNA synthetase is the acylation of ATP by tyrosine to
form tyrosyl adenylate. This reaction can be carried out
within the crystal to form crystals of a complex with
tyrosyl adenylate. The complex is chemically stable, but
the crystals tend to crack during formation of the
complex, and these cracks only partially anneal[23].

It is also possible to soak enzyme crystals with
tyrosinyl adenylate, a powerful inhibitor which is a close
analogue of tyrosyl adenylate[5]. Once again, crystal
cracking is a problem, but it was possible to obtain
diffraction data of considerably higher quality with this
system, though the results do not suggest any significant
difference between the mode of binding of the two
molecules, which only differ in the substitution at the
carbonyl carbon.

Soaking of crystals with ATP at concentrations up to
500mM has produced no crystallographic evidence of ATP
binding[24,25]. Attempts were made to bind ATP to the free
enzyme, and in the presence of tyrosinol and tyramine.
This contrasts with the apparent $K_M$ for ATP in the
aminoacylation reaction of the order of 5mM[16].

The hydrogen-bonding interactions made by tyrosyl
adenylate with the enzyme are shown schematically in
Figure 2. There are not more than three hydrogen bonds
between the enzyme and the adenylate, and only one is with
a side-chain which could be altered by protein
engineering. Cys35 makes contact with the ribose O3', but
at 4.2A the S is too remote for a S .. H-O hydrogen bond,
which would in any case be very weak. Thr51 was at one
time thought to make a hydrogen bond to ribose O4', but
the refined distance of 3.6A is too long for a significant
O .. H-O interaction. There are no strongly basic groups
in the vicinity of the phosphate group.

The adenylate is located in the wide outer cleft
which leads to the tyrosine-binding slot, and in the

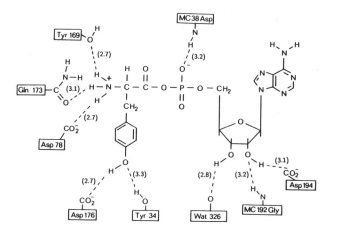

<u>Figure</u> <u>2</u> Polar interactions with significant hydrogen-bonding character between tyrosyl-tRNA synthetase and tyrosyl adenylate. The indicated distances are those between non-hydrogen atoms[2].

adenylate complexes it is well ordered, but the interactions holding it in its observed conformation are weak. Apart from the polar interactions already mentioned, they are weak non-polar interactions.

The adenylate binding site of the enzyme includes much of the structure (residues 34 - 61) found to have an important structural homology with methionyl-tRNA synthetase[26].The sequence homology between <u>B.</u> <u>stearothermophilus</u> tyrosyl-tRNA synthetase and <u>E.</u> <u>coli</u> methionyl-tRNA synthetase in this region is insignificant (5 identities in a span of 13 residues), but it provided the first hint of a much closer series of relationships with the aminoacyl-tRNA synthetases for Met, Ile, Leu, Val, Gln, Glu and Trp whose amino-acid sequences are more closely related to each other in this region than they are to the Tyr enzyme (reviewed elsewhere[2]). Amongst all these enzymes only one glycine (Gly47 of the Tyr enzyme) is fully conserved.


5 PROTEIN ENGINEERING THAT AFFECTS ADENYLATE BINDING

The conformation of Gly47 is such that any side-chain at this position would interfere with the observed adenine conformation. A mutant with alanine at this position has

been constructed and preliminary kinetic measurements
suggest that there is considerable reduction of catalytic
activity[27]. The crystal structure of the mutant[19] showed
extra density for the alanine substitution obtruding into
the adenine site. There was no other structural change in
the adenylate site, apart from a small movement of the
imidazole of His48 produced by contact with the alanine
side chain[28].

The hydroxyl of Thr51 makes only a very weak polar
interaction with the ribose O4', as explained above. Its
main interaction with the adenylate is through contact
with the adenine ring. It is hydrogen bonded to the main
chain O48, and in the absence of adenylate it is exposed
to solvent. A mutant in which this side-chain was altered
to proline was found to have enhanced affinity for
adenylate[14,17]. Brown[19,20,21] investigated the crystal
structures of a series of mutants at position 51, and
showed that the replacement by proline, alanine or serine
side-chains makes no structural change to the enzyme, but
affects the surrounding water structure. This water is
displaced on binding the adenylate, and in the wild-type
enzyme a small vacant void, too small to accommodate a
water molecule, is created adjacent to the threonine
oxygen. Especially in the proline mutant, this void is
neatly filled (Figure 3).

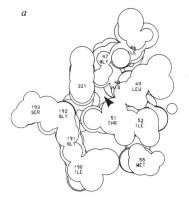

Figure 3 Space filling representations of the adenine-
binding site[20]. Adenine is marked "321". a Wild-type,
with the void adjacent to Thr51 indicated by an arrow. b
Mutant with proline at position 51.

It was shown that the double mutant HG48TP51[13,15] has an effect on the binding energy of the adenylate which is not simply the sum of the effects of the two individual mutations. Brown's results show that this is not due to any structural change in the enzyme introduced by the insertion of proline at position 51, as originally suspected. It may be partly due to the rearrangement of water structure linking the two side-chains, and to the increased solvent accessibility of residue 51 which follows from deletion of the side chain of residue 48.

Brown[21] also studied the crystal structures of the mutants of His48 and Thr51 to glycine. Both showed considerable enhancement in the mobility of the chain in the whole loop linking the mutant residues. This additional mobility may also contribute to the apparent interaction of the two mutations in the double mutant HG48TP51.


6 DISCUSSION

"Improved" enzymes

It was a great surprise that the Pro51 mutant was found to have an enhanced affinity for adenylate[14], since it appeared very unlikely that a simple mutation which enhances the performance of an enzyme would not have taken over as the wild-type. This paradox was resolved when it was shown[11] that the mutant enzyme, though more efficient in forming tyrosyl adenylate, was quite useless in making tyrosyl-tRNA since the tyrosyl adenylate is too strongly bound. Subsequently, other mutants were found at position 51 which stabilise the adenylate just as strongly[21]

In the industrial use of enzymes, it is very common to require them to work under conditions totally different from the natural environment. It seems very likely that protein engineering will provide a valuable tool for improving the performance of every industrial enzyme which has to function in unnatural conditions.

Weak binding

The Pro51 mutant provides an example which shows that weak binding of a ligand is essential for the correct functioning of the enzyme. The weak binding of adenylate has the curious consequence that the binding of ATP has proved impossible to observe crystallographically.

The concentration of ATP in normal cells exceeds $K_M$ for adenylation, so the weak binding of ATP does not impair the operation of the synthetase in vivo. Prior to reaching the transition state, the ATP molecule may take up a wide variety of conformations within the outer cleft. In different conformations, some favourable interactions with the enzyme could be made, but none of them strong enough to fix the ATP conformation. Because of the limited range of conformations accessible to ATP within the cleft, it would lose virtually all of its rotational entropy, much of its conformational entropy and would retain only a little of its translational entropy as it moves around the cleft. Although it is too disordered for its time-averaged electron density to show up clearly by x-ray diffraction, its loss of entropy places it in a state from which the transition state is easily reached. At the transition state, and after the aminoacyladenylate is formed, the adenylate moiety must lose its remaining translational entropy, and becomes fixed in a unique, readily observable conformation.

This picture is consistent with the results on the development of binding energy by the amino-acid residues around the active site cleft, at each stage of the reaction, obtained by Fersht and his colleagues by site-directed mutagenesis[17]. These interactions are very weak for the E.Tyr.ATP pretransition state complex, but become much stronger at the transition state and at the subsequent E.Tyr-AMP.PP$_i$ state. It has been shown that side chains which are considerably distant from the adenylate-binding site can have an important effect on the rate of crossing the transition-state barrier[18].


## 7 CONCLUSION

Protein engineering has introduced a new vista of almost unlimited possibilities in protein chemistry, which are being exploited enthusiastically and rapidly. One encouraging fact shown in our work is that the majority of simple replacements can be made with success, without affecting protein folding and with little or no affect on the local main chain structure. A general exception, which is not too surprising, is that the introduction of glycine confers additional mobility on the local structure.

The crystal structure of a deletion mutant, in which a whole domain was removed[8] also gives a simple result. X-ray crystallography[6] shows the truncated enzyme to have virtually the same structure as the wild-type (apart from the missing domain). There was a small rotation of about $4^o$, of one of the four domains in the truncated enzyme dimer, relative to the other three. Of course the crystal lattice was completely different.

Thus it is not unreasonable to expect that small substitutions, and the excision or shuffling of domains, have a good chance of going ahead as planned, and it may be fair to assume that in the majority of cases the structure will be disturbed to a minimal extent.

The message for designers of protein recognition sites is not quite so simple. In many cases, it may be possible to engineer binding sites with altered specificity on the basis of a simple design. An encouraging recent example is the engineering of a lactate dehydrogenase to be a malate dehydrogenase[29]. But our work has revealed a number of pitfalls.

Recognition is not simply a matter of absolute affinity for a site, but depends on the concentration of competing ligands for the site. It is quite surprising how the recognition systems which have evolved are only marginally good enough, and can be seriously disturbed by a change in the concentration of ligands from that normally found in vivo. There are several examples where the overexpression of a protein has seriously distorted cell metabolism.

Recognition sites should not always be precisely shaped for the desired ligand, nor should their polar interactions be designed solely for favourable interactions when the ligand is fully bound. It is necessary to consider the energetic problems of entering and leaving the site. In particular, the product of an enzyme reaction should not be bound so strongly that its leaving limits the rate of turnover. It is also necessary to remember that if water molecules are in the site, they must be able to leave it as the ligand enters.

Significant effects on binding follow from the water structure around a binding site. Predictions of water structure on enzyme surfaces are not yet being made reliably, and the consequences for binding energies will not be obvious.

It is always tempting to think of a crystallographic-ally known structure as a rigid, static entity. Mobile structures will have a certain "softness" for accommodat-

ing a ligand. Insertion of an amino-acid which increases
the flexibility of a peptide chain (glycine) or decreases
it (proline) will change not only this softness, but also
the entropy of the chain, with consequences for the free
energy changes when a ligand is bound. The results on the
binding of ATP to tyrosyl-tRNA synthetase indicate that
biological recognition does not always require the
complete immobilisation of the ligand at the recognition
site. An incompletely immobilised ligand will possess
extra entropy. If the ligand then forms a chemical bond
to an immobilised atom a significant entropy change
occurs, with consequences for the free energy of the
reaction step.


Acknowledgment. This contribution draws freely on the
work of my colleagues and collaborators, especially Peter
Brick, Katy Brown, Paul Carter, Alan Fersht, Alice
Vrielink, Tim Wells and Greg Winter, not all of which is
fully published. I thank them for their indulgence.


REFERENCES

1.  T.N. Bhat, D.M. Blow, P. Brick and J. Nyborg,
    J.Mol.Biol., 1982, 158, 535.
2.  P. Brick, T.N. Bhat and D.M. Blow, 1989, J.Mol.Biol.,
    in press.
3.  C. Zelwer, J-L. Risler and S. Brunie, 1982,
    J.Mol.Biol., 155, 63.
4.  S. Brunie, P. Mellot, C. Zelwer, J-L. Risler, S.
    Blanquet and G. Fayat, 1987, J.Molecular Graphics, 5,
    18.
5.  C. Monteilhet and D.M. Blow, 1978, J.Mol.Biol., 122,
    407.
6.  P. Brick, and D.M. Blow, 1987, J.Mol.Biol., 194, 287.
7.  S. Brunie, C. Zelwer and J-L. Risler, 1989, Congres
    de Printemps du Societe de Chimie Biologique,
    Grenoble, 6 April 1989. Abstract 8.
8.  M.M.Y. Waye, G. Winter, A.J. Wilkinson and A.R.
    Fersht, 1983, EMBO J., 2, 1827.
9.  A.R. Fersht and M.M. Kaethner, 1976, Biochemistry,
    15, 3342.
10. A.R. Fersht, J-P. Shi, J. Knill-Jones, D.M. Lowe,
    A.J. Wilkinson, D.M. Blow, P. Brick, P. Carter,
    M.M.Y. Waye and G. Winter, 1985, Nature, 314, 235.
11. G. Winter, A.R. Fersht, A.J. Wilkinson, M. Zoller and
    M. Smith, 1982, Nature, 299, 756.
12. A.J. Wilkinson, A.R. Fersht, D.M. Blow and G. Winter,
    1983, Biochemistry, 22, 3581.

13. P.J. Carter, G. Winter, A.J. Wilkinson and A.R.
    Fersht, 1984, Cell, 38, 835.
14. A.J. Wilkinson, A.R. Fersht, D.M. Blow, P.Carter and
    G. Winter, 1984, Nature, 307, 187.
15. D.M. Lowe, A.R. Fersht, A.J. Wilkinson, P. Carter and
    G. Winter, 1985, Biochemistry, 24, 5106.
16. T.N.C. Wells and A.R. Fersht, 1985, Nature, 316, 656,
17. C.K. Ho and A.R.Fersht, 1986, Biochemistry, 25, 1891.
18. A.R. Fersht, J.W. Knill-Jones, H. Bedouelle and G.
    Winter, 1988, Biochemistry, 27, 1581.
19. K.A. Brown, A. Vrielink and D.M. Blow, 1986,
    Biochem.Soc.Trans., 14, 1228.
20. K.A. Brown, P. Brick and D.M. Blow, 1987, Nature,
    326, 416.
21. K.A. Brown, 1988, Ph.D. dissertation, Univ. of
    London.
22. M. Fothergill, 1989, Ph.D. dissertation, Univ. of
    London.
23. J. Rubin and D.M. Blow, 1981, J.Mol.Biol., 145, 480.
24. C. Monteilhet, P. Brick and D.M. Blow, 1984,
    J.Mol.Biol., 173, 477.
25. P. Brick and S. Sundaresan, unpublished work.
26. D.M. Blow, T.N. Bhat, A. Metcalfe, J-L. Risler, S.
    Brunie and C. Zelwer, 1983, J.Mol.Biol., 171, 571.
27. T.N.C. Wells and G. Winter, personal communication.
28. A. Vrielink, personal communication.
29. H.M. Wilks, K.W. Hart, R. Feeney, C.R. Dunn, H.
    Muirhead, W.N. Chia, D.A. Barstow, T. Atkinson, A.R.
    Clarke and J.J. Holbrook 1988, Science, 242, 1547.

# CAVEAT: A Program to Facilitate the Structure-Derived Design of Biologically Active Molecules

By Paul A. Bartlett, Gary T. Shea, Stephen J. Telfer, and Scott Waterman

DEPARTMENT OF CHEMISTRY, UNIVERSITY OF CALIFORNIA, BERKELEY, CALIFORNIA 94720, USA

## 1 INTRODUCTION

The design of enzyme inhibitors is a well developed domain within the fields of bioorganic and medicinal chemistry. The concepts of transition state analogy[1] and suicide inhibition,[2] coupled with knowledge of substrate specificity and enzymatic mechanism, have proven to be very effective for the design of potent compounds. While these "mechanism-derived" approaches do not require any information on the structure of the enzyme, there remain a number of systems for which their application is not straightforward. With the increasing availability of protein structural information, via X-ray crystallography or 2D NMR techniques,[3] a complementary approach to the design of inhibitors (and other types of active compounds) becomes increasingly attractive. A "structure-derived" approach is intuitively logical, since small molecules that are complementary spatially and electronically to an enzyme active site should bind with high affinity. Similarly, a small molecule which imitates the "active loop"[4] of a macromolecular ligand may mimic its binding properties as well. The problem with such structure-derived approaches to inhibitor design is our inexperience as to how actually to carry them out.

A number of examples of *de novo* design of enzyme inhibitors and mimics of macromolecular ligands are known, reflecting varying success.[5] These examples reveal the considerable intuition that the chemist utilizes in deciding what molecule(s) to make and the subjective manner in which the problem is approached. Would the same targets have been designed if a different chemist had tackled the problem? Or the same chemist two months later? To remedy this situation, several automated methods have been developed to help the chemist tackle such design problems in an objective manner.[6-8] The program DOCK[7] allows one to identify molecular structures which fit a binding cavity spatially, and the program GRID[8] provides an indication of what functional groups complement a protein surface electronically. Automated methods of this type are likely to provide a degree of reproducibility in the design process, and to stimulate the chemist to think of structures or groups and to pursue conceptual leads which might otherwise be overlooked.

In parallel with these developments, and indeed stimulated by them, a strong need has arisen for an automated approach to the design of the molecular framework that is to hold the eventual candidate together and fix the functional groups in their proper positions. This report provides an example of a situation in which this need was manifested and a description of the solution we devised.

The 74-amino acid inhibitor of α-amylase known as tendamistat is a particularly attractive target for the design of a peptide structural mimic. The conformation of this molecule has been determined in solution by 2D NMR techniques[9] and in the crystalline state as the isolated inhibitor.[10] There is strong evidence both for the involvement of the triad of residues $Trp_{18}$-$Arg_{19}$-$Tyr_{20}$ as key groups in the active loop of the inhibitor,[11] as well as for a well-defined conformation for the peptide backbone in this region (Figure 1[12]).

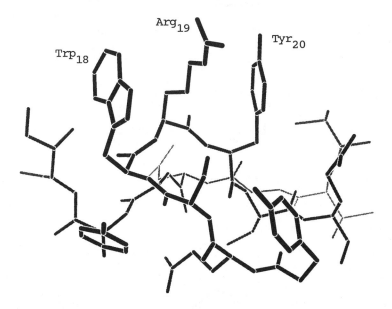

Figure 1    Conformation of tendamistat in the region of the Trp-Arg-Tyr triad; from the crystal structure by Pflugrath, et al.[10]

The orientations of the 3-indolylmethyl, 3-guanidinopropyl, and 4-hydroxybenzyl side chains of residues 18-20 in tendamistat are well-defined in the crystal,[10] although they are likely to be conformationally mobile in solution[9] and may even adopt a different conformation in the enzyme-inhibitor complex. Although one could consider designing a mimic in which a phenolic hydroxyl, a guanidinium moiety, and a hydrophobic indolyl substituent are fixed in the orientations shown in Figure 1, one runs the risk of mimicking a conformation different from that which is bound to the enzyme. An alternative approach would be to design a template which would hold the $C_\alpha$-$C_\beta$ bonds of the side chains in a relative orientation similar to that found in tendamistat, since this relationship appears to be conserved by the inhibitor under a variety of conditions.

The first step in the design process was therefore to identify a conformationally fixed, presumably polycyclic molecule with substituent bonds which adopt the same vector relationship as that between the $C_\alpha$-$C_\beta$ bonds of residues 18-20 in tendamistat. Replacement of the matching bonds with the appropriate side chains could therefore lead to the design of a tendamistat mimic, as suggested schematically in Figure 2.

We found the discovery of the rigid template to be the most subjective aspect to the entire process. Highly effective minimization routines and display and comparison programs were available to evaluate the structures of any templates that we might design, but the actual identification of candidates remained an arbitrary process. We were thus stimulated to automate this step, by developing a program that enables us to search a structural database for molecules whose substituents already match the desired vector relationship. We anticipated that we would uncover different ring systems, and therefore come up with different candidate mimics, than we would have thought of in the relatively haphazard, subjective process.

## 2  DEVELOPMENT OF CAVEAT

The most important database of small molecule structures is the Cambridge Structural Database (CSD).[13] In addition to information on atom type, bond connectivity, and unit cell parameters, the three-dimensional coordinates for each atom are contained in the CSD files. This database can be searched according to a number of criteria, both three-dimensional as well as topological, but it cannot be searched in a manner that would solve the problem outlined above, i.e., to identify molecules that have bonds from three ring atoms to three non-ring atoms which match the three $C_\alpha$-$C_\beta$ bonds of tendamistat. In order to carry out such a search, we have developed the program CAVEAT.[14]

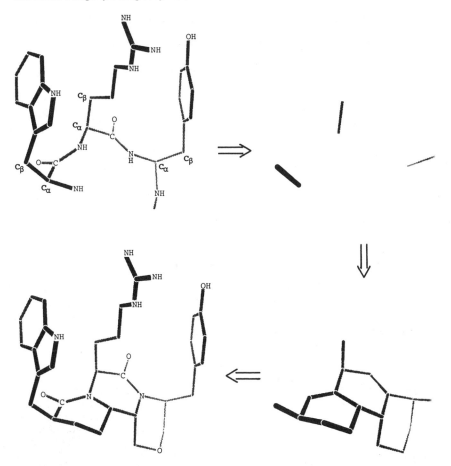

Figure 2   Conceptual steps in the design of a rigid mimic of residues 18-20 in tendamistat.

The foundation of CAVEAT is a database (VBASE) in which the molecules are characterized by the relationships between the substituent bonds, rather than the atomic coordinates of the separate atoms. A substituent bond is defined as a bond between an atom in a ring system (the base atom, B) and an atom not part of the same ring system (the tip atom, T). This definition was adopted to exclude bonds which might be conformationally mobile. Bonds to hydrogen are included in this definition because, although hydrogens are not normally counted as "substituents", they could be replaced with the target side chain.

Each substituent bond is considered a <u>vector</u>, of unit length and with an orientation defined by the base and tip atoms; a <u>vector-pair</u> is defined by two pairs of base and tip atoms, $(B_1, T_1)$ and $(B_2, T_2)$. The relationship between any pair of substituents is defined uniquely by four parameters: $d$, $\delta$, $\alpha_1$, and $\alpha_2$, as illustrated in Figure 3.

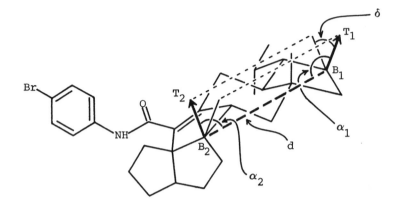

<u>Figure 3</u>    The four parameters that define a vector-pair.

The distance $d$ is the length of the line segment connecting the base atoms of the vector-pair. For our initial evaluation, the range of values for this parameter extended from 3.0 to 9.0 Å. The lower limit of 3.0 Å was chosen to exclude relationships between vectors on the same five- or six-membered ring, for example, which would constitute a highly redundant set; the upper limit was chosen to span most of the organic molecules in the database. The distribution of values for $d$ within this range is illustrated in Figure 4(a), which was obtained from processing a subset of ca. 6000 structures from the CSD.

The dihedral angle $\delta$ is chosen as the absolute value of the lesser of the two dihedral angles between the planes defined by atoms $B_1-B_2-T_2$ and atoms $B_2-B_1-T_1$ (Figure 3). Positive and negative dihedral angles are not differentiated, hence the enantiomer of the pair and the pair itself are indistinguishable. The first occluded angle $\alpha_1$ is the angle between the lines $B_1-T_1$ and $B_1-B_2$; the second, $\alpha_2$, is that between $B_2-T_2$ and $B_2-B_1$. The orientation of the vector-pair is chosen such that $\alpha_1 \leq \alpha_2$. The range of the angle parameters is $0-\pi$ radians; the distribution of values for $\delta$ and $\alpha_1$ within this range is illustrated in Figures 4(b) and (c).

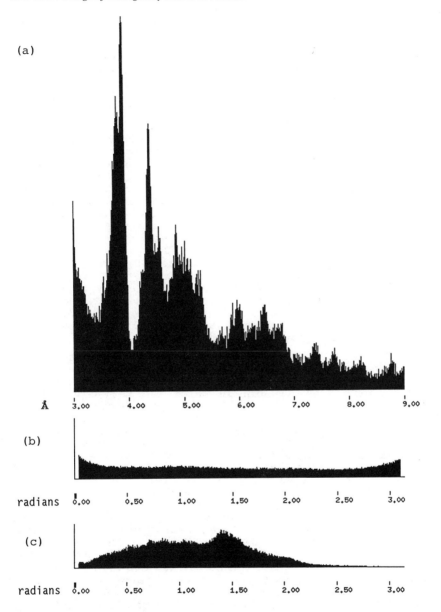

Figure 4   Distribution of values of (a) d (between 3 and 9 Å),
(b) δ, and (c) $α_1$ for a 6,000-molecule sample of the
Cambridge Structural Database.

In order to facilitate the search process, the values that these parameters can adopt are made discrete instead of continuous. The range of values for d, for example, is partitioned into 32 intervals or <u>bins</u>; the angle parameter ranges are similarly subdivided. The number of discrete vector pair relationships is therefore $32 \cdot 32 \cdot 32 \cdot (32+1)/2$ (the latter factor resulting from the condition that $\alpha_1 \leq \alpha_2$). Both the range and distribution of bins can be specified for each parameter by the user.

To create the database, the program VPREP takes as input the ranges and partitions for the four parameters, as specified by the user, and the atom-coordinate and connectivity files for the portion of the CSD to be processed. For each molecule, hydrogen atoms are added if necessary, using a standard geometrical algorithm, and the valid substituent bonds and then vector-pair relationships are identified. For each vector pair, the four parameters are computed and the bin assignment is made, and the data is then incorporated into the database VBASE. This database consists of a list of all the vector-pairs that have been extracted from the CSD; each record indicates the molecule from which the vector-pair was obtained, the four atoms that comprise it, and the bin to which it belongs. The output from VPREP is an open database in the sense that more data may be added to it. When the database is complete, a utility program VPRSS sorts it to improve search speed and compacts it to save storage space.

To search VBASE, the user supplies the coordinates of a vector-pair to be matched, along with an indication of how loose a match will be accepted. The program VSRCH takes this input, identifies the bin in which the target vector-pair falls, as well as any neighboring bins which are included within the tolerance limits, and then provides as output a list of the CSD molecule codenames (REFcodes) and the atoms which comprise the matched vector-pair. To search for a triplet of vectors (a "triple search"), as required for the tendamistat mimic proposed above, the input file is similar except that it contains the coordinates for six atoms, in this case the coordinates of the $C_\alpha$ and $C_\beta$ atoms of residues 18-20 in tendamistat. A triple search is carried out as three pair searches (i.e., the 1-2 pair, the 2-3 pair, and the 1-3 pair), and the hit lists from the three searches are correlated to identify molecules that appear on all three lists and for which the three vector-pairs are defined by only six atoms. Currently, only vector-pair and -triple searches can be carried out with CAVEAT. The searches are relatively fast: a pair search on a database derived from a 6000-molecule sample of the CSD requires less than three seconds on a VAXstation 3200; the time required for a triple search depends strongly on how broad the tolerance limits are, but it is typically less than three minutes.

Because positive values are not distinguished from negative values for the dihedral angle parameter, enantiomeric relationships cannot be distinguished at the vector-pair level. Nor can they be distinguished in a triple search. This has the desirable result of doubling the effectiveness of the database, since a hit whose substituent bonds are enantiomeric to the target is equally valid as a suggestion to the designing chemist.

To evaluate the molecules identified in the search, the structures are retrieved from the CSD using the software provided with it. The CSD atom-coordinate and connectivity files are processed again by VPREP to add hydrogen atoms if necessary, to transform the coordinates into the Cartesian coordinate system, and to write the file in a format (such as Brookhaven or Macro-Model) which can be read by standard molecular graphics programs. An additional utility program TWIST has been developed to transform the coordinates of the candidate structures so that when they are called up on a computer graphics system, they appear with the matching bonds superimposed on the target structure. The various steps of a CAVEAT search are summarized in Figure 5.

## 3 DESIGN APPLICATIONS OF <u>CAVEAT</u>

### Tendamistat

What does CAVEAT suggest as a template for a tendamistat mimic? A few of the molecules retrieved from the CSD are depicted in Figure 6.

Some of these hits, such as the macrocycle DAKREM[15] and the fulvalene trimer BEDFAR[16] are clearly impractical as templates and are not considered further. While the steroid BOCHIK[17] has a number of superfluous rings and side chains, it suggests a simpler tricyclic template that could prove workable. However, further consideration of this template reveals that if the highlighted bonds were replaced with the appropriate side chains, the latter would adopt quite different conformations than are likely to be present in tendamistat. Perhaps the most interesting hit is VHVIMH10, a cyclic hexadepsipeptide which is the D,L,L,L,L,L-isomer of enniatin B.[18] Not only do the three pairs of $C_\alpha$-$C_\beta$ bonds match, there is also a remarkable congruence between the backbone and side chain conformations (Figure 7). The corresponding cyclodepsipeptide containing N-methyl-D-tryptophan, the alcohol analog of L-arginine, and N-methyl-L-tyrosine is an intriguing candidate as a mimic of the structure and action of tendamistat.

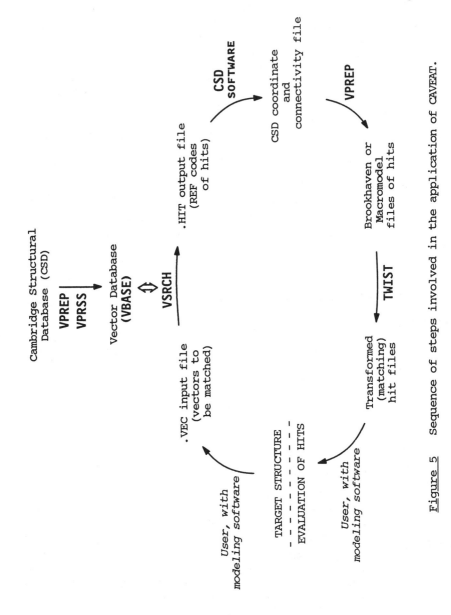

<u>Figure 5</u>    Sequence of steps involved in the application of CAVEAT.

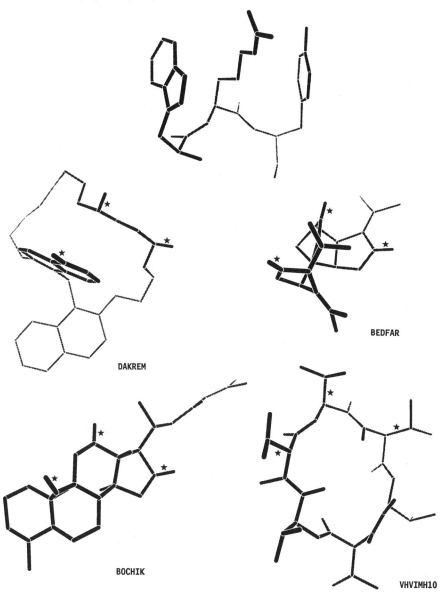

<u>Figure 6</u>   Some CAVEAT hits for tendamistat. The starred bonds are those that match the $C_\alpha$-$C_\beta$ bonds of residues 18–20 of the target.

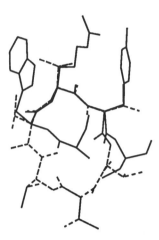 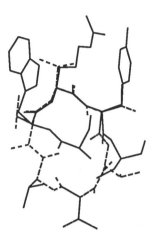

Figure 7   Comparison between enniatin B and residues 18-20 of
tendamistat (convergent stereopair).

Somatostatin

     In what is now a classic series of investigations, a
number of cyclic peptides were developed as analogs of somatostatin by Freidinger and Veber and their colleagues at Merck.[19]
One of the most potent analogs, cyclo(Pro-Phe-D-Trp-Lys-Thr-Phe),
was studied extensively by NMR and computational techniques to
provide insight into the conformation of the molecule and the
segment responsible for receptor recognition.  As an exercise
in the application of CAVEAT to the design of a non-peptidic
mimic, we selected as a target one of the low energy conformations of this molecule, as predicted by the Merck group.  The
three $C_\alpha$-$C_\beta$ bonds of residues L-Phe-D-Trp-L-Lys were chosen as
input to CAVEAT.  One of the hits, BXDABT,[20] an abietic acid
derivative (Figure 8), was then carried through several cycles
of modification, minimization, and comparison, to find structures
which matched better and which might be easier to synthesize.
In Figure 8 are displayed the structures of the initial hit, the
modified mimic, and a superposition of the latter with the
target somatostatin analog.  Although further modeling and
evaluation would be warranted before synthesis of such a molecule
were undertaken, the fact that the entire sequence, including
the CAVEAT search, was carried out in an afternoon indicates
that the design process is fairly efficient.

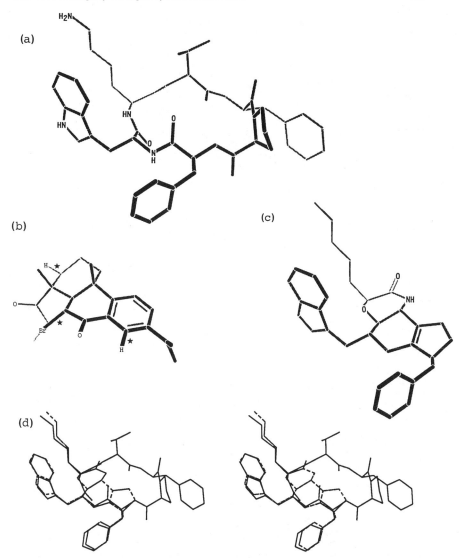

Figure 8    (a) Predicted low energy conformation of the somatostatin analog cyclo(Pro-Phe-D-Trp-Lys-Thr-Phe),[19] with
the search vectors highlighted;    (b) BXDABT (methyl
6-bromo-7-oxodehydroabietate)[20] with the matching
bonds highlighted;    (c) a potential mimic derived
from from BXDABT;    (d) overlap of the potential
mimic (dashed lines) with the somatostatin analog
(solid lines) (convergent stereopair).

Application in *de novo* inhibitor design

Can CAVEAT be used in the absence of something to mimic, for example if the structure of the binding site, but not that of a ligand, is available? As noted in the introduction, there are a number of programs available which tell one what groups should be placed in a binding site and even where they should be placed, but not how they should be connected. However, asking how groups should be connected is simply another way of asking for a template which provides a given vector relationship. We therefore envisage that CAVEAT could be used in concert with other programs to address the issue of *de novo* design as well.

## 4 CAVEATS ABOUT CAVEAT

It is a tenet of bioorganic chemistry that conformational flexibility reduces binding affinity, other things being equal. The design of peptide mimics such as those described above rests on this foundation, although the quantitative significance of this factor remains largely unexplored. Also poorly understood is the role that conformational flexibility plays in the kinetic, as opposed to thermodynamic, aspects of binding. The underlying validity of the approaches outlined above, the fixing of side chain groups on rigid templates, has therefore not yet been demonstrated.

While in each instance the accuracy with which the target structure has been determined can be assessed, how tolerant the binding process is to deviations or misplacements of the side chains in the mimic cannot be predicted yet with confidence. It must be recognized that the "resolution" of a CAVEAT search will in most instances be much better than is warranted by these two factors. A CAVEAT search therefore cannot be exhaustive, and its value must be recognized for the diverse templates or ring systems that it suggests to the user and which might not otherwise have been recognized.

## 5 SYSTEM REQUIREMENTS AND AVAILABILITY OF CAVEAT

CAVEAT was written in C and is available for computers running UNIX or VMS operating systems;[21] at present, use of this program requires access to the Cambridge Structural Database.[22] CAVEAT is licensed by the University of California; further information is available from P.A. Bartlett.

**REFERENCES**

1. R. Wolfenden, <u>Annu. Rev. Biochem. Bioeng.</u>, 1976, <u>5</u>, 271. R. Wolfenden and L. Frick, 'Enzyme Mechanisms', M.I. Page and A. Williams, eds., Royal Society of Chemistry, London, 1987, p. 97.

2. R.H. Abeles, <u>Pure Appl. Chem.</u>, 1980, <u>53</u>, 149. C. Walsh, <u>Tetrahedron</u>, 1982, <u>38</u>, 871.

3. K. Wüthrich, <u>Science</u>, 1989, <u>243</u>, 45; <u>Accts. Chem. Res.</u>, 1989, <u>22</u>, 36.

4. We use the term "active loop" to refer to the binding region of a convex surface in the same sense that "active site" refers to the binding region of a concave surface.

5. <u>Inter alia</u>: H.L. Sham, G. Bolis, H.H. Stein, S.W. Fesik, P.A. Marcotte, J.J. Plattner, C.A. Rempel, and J. Greer, <u>J. Med. Chem.</u>, 1988, <u>31</u>, 284. W.C. Ripka, W.J. Sipio, and J.M. Blaney, <u>Lect. Heterocyclic Chem.</u>, 1987, <u>IX</u>, 95. L. Sheh, M. Mokotoff, and D.J. Abraham, <u>Int. J. Peptide Res.</u>, 1987, <u>29</u>, 509. M.G. Bock, R.M. DiPardo, K.E. Rittle, D.E. Evans, R.M. Freidinger, D.F. Veber, R.S.L. Chang, T.-b. Chen, M.E. Keegan, and V.J. Lotti, <u>J. Med. Chem.</u>, 1986, <u>29</u>, 1941. M. Mutter, K.-H. Altmann, K. Müller, S. Vuilleumier, and T. Vorherr, <u>Helv. Chim. Acta</u>, 1986, <u>69</u>, 985. C. Freudenreich, J.-P. Samana, and J.-F. Biellmann, <u>J. Am. Chem. Soc.</u>, 1984, <u>106</u>, 3344.

6. Y.C. Martin, E.B. Danaher, C.S. May, and D. Weininger, <u>J. Computer.-Aided Mol. Des.</u>, 1988, <u>2</u>, 15. S. Namasivayam and P.M. Dean, <u>J. Mol. Graphics</u>, 1986, <u>4</u>, 46.

7. R.L. DesJarlais, R.P. Sheridan, G.L. Seibel, J.S. Dixon, I.D. Kuntz, and R. Venkataraghavan, <u>J. Med. Chem.</u>, 1988, <u>31</u>, 722.

8. P.J. Goodford, <u>J. Med. Chem.</u>, 1985, <u>28</u>, 849.

9. A.D. Kline, W. Braun, W., and K. Wüthrich, <u>J. Mol. Biol.</u>, 1986, <u>189</u>, 377; <u>ibid.</u>, 1988, <u>204</u>, 765. See also Reference 3.

10. J.W. Pflugrath, G. Wiegand, R. Huber, and L. Vértesy, <u>J. Mol. Biol.</u>, 1986, <u>189</u>, 383.

11. O. Epp and R. Huber, personal communication.

12. The structures in this report were generated using the program SHADEMOLE: M. Hahn and W.T. Wipke, Tetrahedron Comput. Methodol., 1988, 1, 81.

13. F.H. Allen, O. Kennard, W.D.S. Motherwell, W.G. Town, and D.G. Watson, J. Chem. Doc., 1973, 13, 119.

14. The acronym derived from "Computer Assisted Vector Evaluation And Target design" also reflects our belief that users of design programs should maintain a healthy skepticism.

15. W. Clegg, J.C. Lockhart, and M.B. McDonnell, J. Chem. Soc., Perkin 1, 1985, 1019.

16. B. Ubersax, M. Neuenschwander, and R. Engel, Helv. Chim. Acta, 1982, 65, 89.

17. V.J. Paul, W. Fenical, S. Rafii, and J. Clardy, Tetrahedron Lett., 1982, 23, 3459.

18. T.G. Shishova and V.I. Simonov, Kristallografiya, 1977, 22, 515.

19. R.M. Freidinger and D.F. Veber, ACS Symp. Ser., 1984, 251 (Conform. Directed Drug Des.), 169.

20. J.F. Cutfield, T.N. Waters, and G.R. Clark, J. Chem. Soc., Perkin 2, 1974, 150.

21. Copyright acknowledgements: VMS is a trademark of the Digital Equipment Corporation; UNIX is a trademark of AT&T.

22. Information on the Cambridge Structural Database is available from the Cambridge Crystallographic Data Centre, University Chemical Laboratory, Lensfield Road, Cambridge CB2 1EW, England.

# Host–Guest Complexation in Organic Solvents. Experimental and Theoretical Results

By W. Clark Still, Jeremy Kilburn, and Philip Sanderson

DEPARTMENT OF CHEMISTRY, COLUMBIA UNIVERSITY, NEW YORK, NY 10027, USA

A primary goal of our research program is the development of methods for designing new molecules with selective binding properties. We approach the problem using a combination of theory and experiment. The theory we use is based on molecular mechanics and is directed toward predicting binding free energies. We use experiment to synthesize novel host structures and to establish their three-dimensional structures and binding properties. In this lecture, I will describe studies of macrotricyclic host molecules which bind guest molecules having hydrogen bond donor and acceptor functionalities. I will try to convey what we have learned about the nature of our host/guest complexes and the forces which hold them together.

## Experimental Results.

A Meso Host Molecule.

Shown below is a cartoon of the type of host/guest (H/G) complex we wish to make along with the structure of a corresponding host. Since the complex is to form in organic solvents, formation of hydrogen bonds provides the main driving force for complexation. In the cartoon, the '+' and '-' represent hydrogen bond donors and acceptors respectively. In the host structure, the hydrogen bond donors are the amide N-H's and the hydrogen bond acceptor is the carbonyl of the cyclic urea.

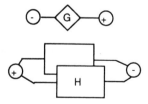
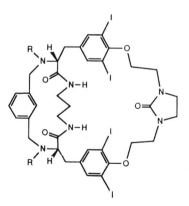

To prevent the hydrogen bond donor and acceptors in the host from forming intramolecular hydrogen bonds which would disfavor binding, we connected these functionalities by *para*-substituted aromatic spacers. To orient these aromatics tangentially with respect to the binding cavity, we provided the rings with two bulky *ortho* substituents (iodines) which force the aromatic ether linkages to be perpendicular to the plane of the aromatic ring. Finally we provided a *meta*-disubstituted aromatic spacer to further restrict the conformational freedom of the host. Thus, we have gone to considerable lengths to reduce conformational heterogeneity and to favor structures having open binding cavities.

The synthesis of the host described above was carried out starting from tyrosine, N,N-diethanol ethyleneurea and *meta*-xylene dibromide. I would like to add only that the Mitsunobu coupling of the diiodotyrosine fragments with primary alcohols is an

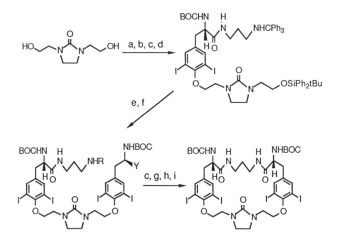

effective and high yield procedure for joining these amino acid sidechains to other organic functionalities. As such, it may be of substantial utility in other constructions of conformationally restricted molecules having large binding cavities.

As summarized in the diagram below, our host does indeed bind small molecule organic guests which have appropriately oriented hydrogen bond donating and accepting regions. It should be noted that only guests having both donating and accepting functionalities bind detectably. Thus imidazole and derivatives having an N-H bind with association energies of ~3-5 kcal/mol. However, if either the donating

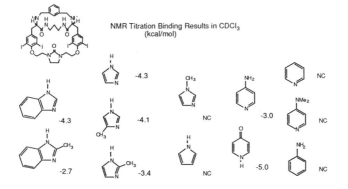

NMR Titration Binding Results in CDCl₃
(kcal/mol)

or accepting functionality is removed (as in N-methylimidazole or pyrrole), binding is lost ('NC' in the diagram above indicates no observed complexation). This result is compatible with the idea that the binding enthalpy is on the order of 10 kcal/mol and that the loss of translational and rotational entropy brings the binding free energy to the observed ~4 kcal/mol. The 10 kcal/mol corresponds approximately to the energy of two hydrogen bonds. In the case of complexes held together by only one hydrogen bond, the favorable binding enthalpy of ~5 kcal/mol is opposed by an unfavorable binding entropy which is of comparable or greater magnitude. Thus the first hydrogen bond may be thought of as offsetting the loss of translational and rotational entropy during binding and the second providing the observed binding energy. This analysis would suggest that adding a third hydrogen bond would bring the binding energy to 10 kcal/mol or more.

Using x-ray crystallography and NMR, we have learned a great deal about the structures of our host and its complexes. In the diagrams below, I have provided stereopair plots of the x-ray crystal structures of the host and its N-benzyl derivative.

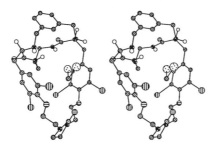

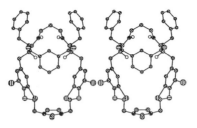

These structures reveal some interesting properties of the host structure. First, there are several low energy conformations of the macrotricyclic ring system. One of the most obvious distinctions is the presence or absence of bridgehead (alpha) hydrogens which point into or out of the binding cavity. Second, the hydrogen bond accepting urea does not point directly into the cavity in these structures. Third, the x-ray structures all show a large, open binding cavity. It is occupied by a methylene chloride (solvent) molecule in one of the structures.

While we have been unable to obtain crystals of the host/guest complexes, we have studied the host/4-pyridone complex by NMR. As shown in the diagram below, a number of nOe signals may be observed which characterize the conformation of the host and the orientation of the guest within the binding cavity. The host conformation for the complex in solution is similar to the x-ray structure of the N-benzyl host above (bridgehead hydrogens point into the cavity). Intramolecular nOe's show that the guest 4-pyridone aligns itself as expected to donate a hydrogen bond to the urea carbonyl and accept a hydrogen bond from one (or both) of the amide N-H's. Indeed, donor hydrogens in both the guest and host show large (1-2 ppm) downfield chemical shifts upon complexation in $CDCl_3$.

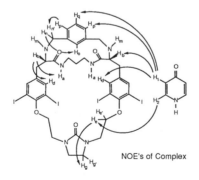

NOE's of Complex

The above data is compatible with a structure for the complex in which the donor/acceptor guest is encapsulated within the host structure in the conformation observed for the benzyl derivative.

A Chiral Host Molecule.

Before turning to theoretical results, we would like to outline one more study on host/guest systems of the type described above. This new study differs however in that it is concerned with enantioselection.

Since the hosts above are prepared from the naturally occurring amino acid tyrosine, there is an opportunity for converting the meso host design into a chiral, enantiomerically pure one. This may be done most easily by constructing a related C2 host. An example of such a structure is shown below:

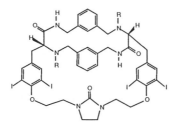

The preparation of this C2 host is straightforward as the synthesis below shows. Thus, beginning from our urea diethanol, we can construct both sides of the binding site simultaneously from L-tyrosine. The double macrocyclization is straightforward and yields the desired chiral host in moderate yield.

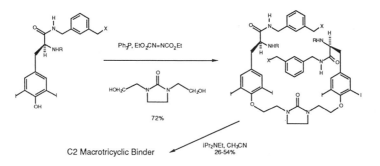

We have obtained several x-ray structures of our C2 host and its derivatives and we again find that there are two major conformers. As before, these differ primarily by the presence or absence of bridgehead hydrogens which point into the binding cavity. The x-ray structure of the simple C2 host is shown below.

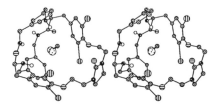

Like the meso host described previously, the C2 host binds hydrogen bond donor/acceptor guests such as imidazole. The binding energies in $CDCl_3$ are similar in magnitude to those listed above. To assess the chiral binding properties of our new host, we looked at another type of donor/acceptor guest, a simple carboxylic amide. The binding constants with simple amides such as N-methylacetamide were small and difficult to measure in $CDCl_3$. However, by going to a less polar solvent (e.g. $C_6D_6$), we found binding energies on the order of 2-3 kcal/mol. The binding energies ($\Delta G$) for several simple amides in $C_6D_6$ are given below:

| | |
|---|---|
| $CH_3CONHCH_3$ | -3.2 kcal/mol |
| $HCONHCH_3$ | -3.7 kcal/mol |
| $CH_3CONHCH_2C_6H_5$ | -2.8 kcal/mol |
| $C_6H_5CH_2CONHCH_3$ | -2.2 kcal/mol |

We next examined the binding of our C2 host to amides of several chiral amines. Shown below is a section of the nmr spectrum of the complex with racemic N-formyl phenylethylamine. The diastereomeric complexes of the enantiomeric amides are clearly distinguished.

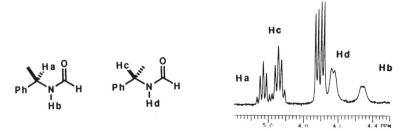

Upon measuring the binding energies of our host with the enantiomers of formyl and acetyl phenylethylamine, we found that the enantiomeric guests were also distinguished by differences in binding energy. As shown below, the S enantiomer bound more tightly in both instances. The acetamides of R and S alanine benzyl ester showed a somewhat greater difference in binding energies:

Binding Energies of C2 Host with Chiral Amides ($C_6D_6$)

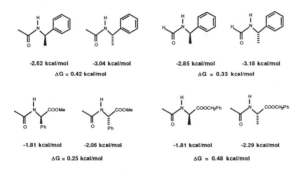

| -2.62 kcal/mol | -3.04 kcal/mol | -2.85 kcal/mol | -3.18 kcal/mol |
| $\Delta G = 0.42$ kcal/mol | | $\Delta G = 0.33$ kcal/mol | |

| -1.81 kcal/mol | -2.06 kcal/mol | -1.81 kcal/mol | -2.29 kcal/mol |
| $\Delta G = 0.25$ kcal/mol | | $\Delta G = 0.48$ kcal/mol | |

While these energy differences are not large, they are experimentally significant. The nmr titrations which were used to measure the binding energies were conducted to >80% saturation and an error propagation analysis gave error limits on the order of 0.1 kcal/mol. To improve enantioselectivity, it will be necessary to learn more about the nature of the selection. To this end we are now studying the relative binding energies computationally. At this point, we are concerned that the observed ~0.5 kcal/mol free energy difference is too small to be able to extract from a calculation. We may need to find a more selective system empirically before we can use molecular modeling to understand relative binding well enough to predict the properties of closely related new hosts.

## Theoretical Results

Our purpose for conducting the experiments described above was to obtain firm data with which to test methods for calculation of binding energies. Although the elementary forces which control binding are well understood, calculating a binding constant is no easy matter. There are many factors which need to be considered - the entropy loss which accompanies binding, the effect of solvation energy, the precise three-dimensional structure(s) of the host/guest complex.

Some of these problems are easier to handle than others. The question of the three-dimensional structure is a particularly thorny problem. If one does not know the lowest energy conformation(s) of the molecules involved, then there is little chance of predicting binding properties. In principle, one could search conformational space to find all conformers and compute their populations using molecular mechanics. However, for flexible structures, there may be so many conformations, and the accuracy of molecular mechanics energies is so poor that finding and distinguishing the populated conformers is problematic. Presently, the best approach is to choose hosts which simplify the conformational problem. Thus, we design molecules which are geometrically well-defined by the presence of conformationally restrictive bonding arrays (macro rings, *para*-substituted aromatics, etc). We further attempt to determine the geometrical properties of structures by x-ray and/or nmr analysis once they are prepared.

Evaluating Solvation Energy.

Solvation energies may be calculated in a number of ways. The most established method is to surround the solute (or the part of it which is of interest) with several hundred solvent molecules. Then, molecular dynamics or Monte Carlo methods can be used to sample a fraction of the available solvent configurations to yield average solvation energies. Coupled with statistical perturbation theory, such simulations yield precise estimates of solvation free energy differences. The main problem with such methods is that they are very time-consuming. Since our objective is to compute binding free energies for the purpose of designing new host molecules, we need a faster approach.

To estimate solvation energies, we have been developing a method which treats the solvent medium as a statistical continuum. This solvent continuum is bounded by the surface of solute but is otherwise infinite in dimensions. It has energetic properties which mimic those of the actual solvent:

    a. A Cavity Term - the energy cost for introducing a hole of molecular dimensions into the solvent continuum.

    b. A Dispersion Term - the van der Waals attractions between the atoms of the solute and the solvent continuum.

    c. A Polarization Term - the attractions of the partially charged atoms of the solute and the polarizable solvent medium.

Each of these terms is evaluated using existing models or simple extensions. I will not go into the details of these solvation energy calculations but simply summarize the nature of the models. For the cavity term, we use an established model known as scaled particle theory (*Chem.Rev.*, **76**, 717 (1976)) and base the calculation on the solvent accessible surface area of the solute. For the dispersion term, we integrate the Lennard Jones equation over a continuum solvent having Lennard Jones parameters taken from the solvent composition and density. The integration is performed from the van der Waals surface of the solute to a total distance from the surface of 8Å. To compute polarization, we use two different methods. For small molecules, we use a free energy perturbation method in which an originally uncharged solute in a bath of solvent (here $CHCl_3$) is electrostatically charged. Such FEP calculations were carried out using the BOSS program of Bill Jorgensen. For larger molecules, we use the Langevin dipole method of A. Warshel (*J.Mol.Biol.*, **185**, 389 (1985)). The two methods give similar results with small molecule solutes.

In the examples to follow, we compute the three solvation terms for each species involved in the equilibrium. Since the calculations are numerical, the solvation energies were evaluated on energy minimized structures.

Evaluating Average Entropy and Enthalpy.

When two independent molecules join to form a bimolecular complex, three translational and three rotational degrees of freedom are converted into low frequency vibrational modes in the complex. Other changes occur as well. The moments of inertia and the vibrational frequencies of isolated molecules and complexes are not the same. Energetically, these changes modify entropy and therefore have an effect on the free energy difference which controls the position of an equilibrium.

Traditionally, entropy changes have been evaluated by calculating the moments of inertia and the harmonic vibrational frequencies from normal mode analysis. Using these calculations however is to assume that study of an energy minimized structure (which corresponds to 0 °K) will provide information which is transferable to the system at higher temperatures. For simple molecules whose vibrations are relatively harmonic, this analysis is adequate. However, for large host-guest molecules and their complexes, potential surfaces at 300 °K tend to be very anharmonic. The diagrams below illustrate the differences between the smooth, approximately harmonic potential wells of simple molecules (left) and the irregular wells of complex molecules (right).

While the average enthalpy and entropy of a molecule in a harmonic energy well can be evaluated from the nature of the bottom of the well, it is difficult to obtain reliable information in the more general case of a complex molecule with a complex potential surface.

To obtain energetic information from the various regions of a complex potential well, we use constant temperature molecular dynamics. With this technique, the atoms of a molecule are given random velocities corresponding to the average kinetic energy of the system at the desired temperature (e.g. 300 $^\circ K$). Using the molecular mechanics force field and Newton's second law of motion ($F = mA$), the positions of the atoms may be determined as a function of time. Over the course of such a molecular dynamics simulation, all regions of a potential well are explored and average thermodynamic properties may be evaluated. If there are several potential wells which are separated by large energy barriers (and thus not readily crossed during the simulation), then separate simulations are carried out within each well.

For the purpose of comparing stereoisomers (including conformers) by molecular mechanics, enthalpy (H) is usually taken as the minimum molecular mechanics steric energy. Such a treatment describes a molecule at 0 $^\circ K$. Using molecular dynamics, the average enthalpy (<H>) can be evaluated as the average steric energy found during a simulation at constant temperature. With adequately long simulations (typically 50-250 ps), the same average enthalpy is found regardless of the precise starting geometry used.

To evaluate the average entropy (<S>), we monitor internal coordinates (e.g. bond angles, torsion angles) during the molecular dynamics simulation and analyze the data using the quasiharmonic entropy method of Karplus (*J.Am.Chem.Soc.*, **107**, 6103 (1985)). The basic approach of the method is summarized in the diagram below which shows how differences in vibrational entropy for a single internal coordinate, a bend, can be evaluated during a molecular dynamics simulation.

In the example, we consider the entropy difference between bending vibrations which differ in frequency due to a difference in bending force constant. If the constant is large, then the vibrational frequency will be high and the entropy low. That case corresponds to the narrow energy well on the left in the diagram below. If the force constant is small, then the vibrational frequency will be low and the entropy high. That case is illustrated by the broader energy well on the right:

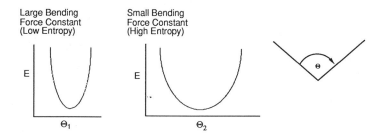

Large Bending
Force Constant
(Low Entropy)

Small Bending
Force Constant
(High Entropy)

Let $\sigma_1$ be the variance of $\Theta_1$ during simulation $(\sigma_1 = <(\Theta_1 - <\Theta_1>)^2 >)$
Let $\sigma_2$ be the variance of $\Theta_2$ during simulation $(\sigma_2 = <(\Theta_2 - <\Theta_2>)^2 >)$

Then $\Delta S = S_1 - S_2 = 0.5\,R\ln(\sigma_1/\sigma_2)$

While the entropy difference ($\Delta S$) could be computed by the normal mode analysis from vibrational frequencies, it can also be evaluated by molecular dynamics simulations of the two different cases. If case 1 (left) were simulated, then one could monitor the bending angle $\Theta_1$ and compute its standard deviation or variance ($\sigma_1$) with respect to the average value of the angle. If case 2 (right) were then simulated, its analogous variance ($\sigma_2$) could be obtained. Since case 2 has a smaller force constant and thus broader energy well than case 1, its variance ($\sigma_2$) would be found to be larger than that of case 1 ($\sigma_1$). The difference in entropy between case 1 and case 2 is given by $0.5\,R\ln(\sigma_1/\sigma_2)$.

The quasiharmonic method outlined above is readily extended to complex systems with many internal coordinates having both correlated and uncorrelated motion. To apply the method to problems of bimolecular complexation, we need additionally to define a set of internal coordinates which relates the position and orientation of the two independent molecules. Such a set of internal coordinates is shown below. Translation is defined as differences in centers-of-mass in the total principal axis frame, and rotation is defined by Euler angles between the principal axes of the individual molecules. These internal coordinates may be monitored like any other internal coordinates in a molecular dynamics simulation. Since the unbound molecules are free to translate and rotate, we know how these distances and angles vary in time.  Free translation corresponds to free movement within a volume defined by the concentration (e.g. a 1M standard state).  Free rotation allows full rotation around each of the three axes.  From free (square well) translation and rotation, we know the corresponding translational and rotational variances with respect to an arbitrary position and orientation.  Thus we can compute the variances for relative translation and rotation for the bound and unbound states of the molecules.

Configurational  Entropy  -  Translation  and  Rotation  Coordinates

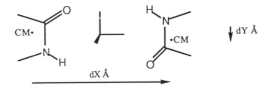

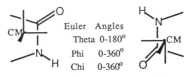

With the methods described above for computing solvation energies and average enthalpy and entropy, we studied the binding properties of a simple association reaction, the dimerization of cis N-methylacetamide.  This molecule was chosen because an experimental free energy of association in chloroform (-0.3 ±0.3 kcal/mol) was available for a closely related cyclic lactam.  We began by energy minimizing two isolated cis N-methylacetamide molecules and the corresponding dimer using Bill Jorgensen's OPLS force field.  Next we carried out 100 ps of molecular dynamics at constant temperature (300 °K) to equilibrate the systems.  Finally, we ran a series of 100 ps simulations in which the average enthalpy and entropy (<H> and <S>) were evaluated.  As summarized below, the reaction is enthalpically favorable in the gas phase to the extent of -12.3 kcal/mol but entropically disfavored by 28.5 eu (T$\Delta$S = +8.5 kcal/mol) given a 1M standard state.  Thus the association free energy is -3.8 (±0.4) kcal/mol *in vacuo*.  Calculating the solvation energies of the free and bound molecules, we find the unbound materials are more favorably solvated than the complex.  The total solvation energy change upon binding is +3.6 kcal/mol (using the FEP solvent polarization energies). Overall, the calculation agrees well with experiment.

2

**Experimental:[1]** $\Delta G_a$ **(CHCl$_3$)** = (-) **0.0-0.6 kcal/mole**

$\Delta G_a$ (CCl$_4$) = (-) 2.7-3.4 kcal/mole

| | | | |
|---|---|---|---|
| $E_{opls/a}$ (dc = 1.0) MM | -28.0 | -42.5 | del = -14.5 kcal/mole |
| 300° MD | -20.7 (±.05) | -33.2 (±.05) | del = -12.3 kcal/mole* |
| $G_{cav}$ (solvation) | +9.6 | +7.9 | del = -1.7 kcal/mole* |
| $E_{vdw}$ (solvation) | -17.2 | -14.0 | del = + 3.2 kcal/mole* |
| $G_{pdld}$ (solvation, av) | -3.2 | -0.5 | del = + 2.7 kcal/mole |
| $G_{fep-chg}$ (solvation, av) | -3.1 | -1.0 | del = + 2.1 kcal/mole* |
| $S_{quasiharmonic}$ | $\Delta S$ = -28.5 (±1.0) cal/deg-mole | | del = + 8.5 kcal/mole* |
| $S_{q-trans,rot}$ | $\Delta S$ = -25.4 cal/deg-mole | | |
| $S_{harmonic}$ | $\Delta S$ = -26.2 cal/deg-mole | | del = + 7.9 kcal/mole |
| | Total Free Energy | $\Delta G_a$ = | **-0.2 (±0.4) kcal/mole*** |

[1]Krikorian, J.Phys.Chem, 1875 (1982) - for corresponding lactam

The main defect of the model is that the molecular dynamics simulations are carried out *in vacuo* and the solvation energies are added afterwards. We have yet to determine how this crude way of handling solvation affects the calculated binding constants. In organic solvents, the errors may not be large because the organic solvation energies are do not change greatly with small changes in geometry and thus may not greatly affect the molecular dynamics simulations.

We have applied the calculational scheme described above to some of our host/guest complexes. In particular, we have computed the free energy of association for our meso host with imidazole in chloroform. These calculations required ~500 ps of molecular dynamics per structure. This length of simulation requires ~5 days on a MicroVAX II or 4 hours on a Convex C210 computer. As shown on the following page, the results match closely with our measured binding free energies. As in the case of cis N-methylacetamide, the binding is highly favorable in the gas phase and has an (unfavorable) entropy of association of ~28 eu. Most of the entropy decrease upon binding comes from the loss of translational and rotational freedom. Solvation energies also work against binding. A similar calculation on the meso host and pyrrole (which do not bind detectably), gave a small binding free energy in chloroform of ~-1 kcal/mol. We are now testing the computational scheme with other host/guest complexes to evaluate the reliability of the calculation.

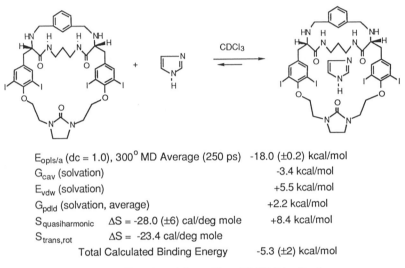

$E_{opls/a}$ (dc = 1.0), $300^{\circ}$ MD Average (250 ps)      -18.0 ($\pm$0.2) kcal/mol
$G_{cav}$ (solvation)                                                              -3.4 kcal/mol
$E_{vdw}$ (solvation)                                                            +5.5 kcal/mol
$G_{pdld}$ (solvation, average)                                             +2.2 kcal/mol
$S_{quasiharmonic}$   $\Delta S$ = -28.0 ($\pm$6) cal/deg mole      +8.4 kcal/mol
$S_{trans,rot}$        $\Delta S$ = -23.4 cal/deg mole
                Total Calculated Binding Energy          -5.3 ($\pm$2) kcal/mol

**Experimental[25]**   $\Delta G$ = -4.3 ($\pm$0.4) kcal/mol
                                    $\Delta S$ = -22.4 ($\pm$6) cal/deg-mol

        In addition to the potential problems of handling solvation separately from the solute free energy calculations, large molecules have a problem of convergence. It is necessary to conduct several long molecular dynamics runs to bring the total precision of the calculation to a useful level of ~1 kcal/mol. While there may be ways to improve the precision without increasing the time, it is likely that the best solution in the near future is faster computing hardware. With a machine having the power of the Convex C210 (speed 30-40 MicroVAX II's), it appears possible to compute binding energies in less than one day. The most recent crop of superworkstations is beginning to offer this level of performance at an almost affordable price.

        In conclusion, we are not far from having the ability to compute reliable binding energies of hypothetical structures in solution for the purpose of designing exciting new molecules. We believe that the availability of reliable predictions will greatly enhance the effectiveness with which research in molecular recognition is carried out.

# Stereoelectronic Effects in Molecular Recognition

Julius Rebek, Jr.

DEPARTMENT OF CHEMISTRY, UNIVERSITY OF PITTSBURGH, PITTSBURGH, PENNSYLVANIA 15260, USA

Bioorganic chemistry deals with the application of organic chemical principles to problems of biological significance and recent events have focused international attention on this area. The award of the Nobel prizes to Pederson, Cram and Lehn in 1987 gave high visibility to the field and lent respectability to the use of model compounds. The use of such molecules to demonstrate binding selectivity, transport and some aspects of regulation has been quite successful. If the field is to grow and to compete with recent developments in the chemistry of antibodies, then similar successes need to be scored in the areas of molecular recognition and catalysis.

The classical molecules of bioorganic chemistry, the macrocyclic compounds, are at a disadvantage with respect to these goals. Functional groups attached to macrocycles tend to diverge away from the substrates held within. That is, they fail to focus catalytically useful functionality on the substrate in the same sense that functional groups converge at the active site of an enzyme or receptor or antibody (Fig. 1). Macrocyclic compounds are also highly specialized for the types of structures that they bind; considerable molecular engineering is required to have substrate come into contact with reactive surfaces on any particular host structure.[1]

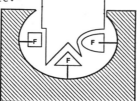

Figure 1   Functional Groups (F) converge to create an active site.

Because of these disadvantages, we have decided to abandon
macrocyclic compounds for more promising shapes, and we have
settled upon a **molecular cleft.** The module from which these
clefts are constructed is the Kemp triacid[2] - a unique structure
in which three carboxyl groups are held in an axial conformation
by dint the larger A values of the equatorial methyl groups. This
conformation insures that a U-shaped relationship exists between
any two carboxyl functions and this structural feature permits the
construction of molecules which fold back upon themselves. The
combination of suitable spacer elements with two Kemp triacid
units yields structures in the general shape of a molecular cleft.
Specifically, mere fusion of two equivalents of the Kemp triacid
and acridine yellow provides a unique structure in which the two
carboxyl OH bonds converge toward the center of the molecule (Fig.
2). Crystallographic studies of this diacid showed that ~ 8.5 Å
separate opposing carboxyl oxygens. In this regard the condensation product shows similarity to the convergent carboxyl functions of, say, lysozyme or the aspartic proteases.

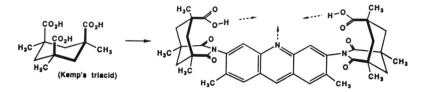

Figure 2   Convergent functionality from the condensation of Kemp's
           triacid with acridine yellow

We have examined the binding of this material to molecules of
complementary size, shape and functionality. For example, heterocyclic diamines to form 1:1 complexes with high selectivity:
DABCO, pyrazine and their derivatives are tightly bound, whereas
stronger bases which lack the appropriate dimensions or functions
are shunned by these molecules.[3]   A typical complex is shown in
Fig. 3.

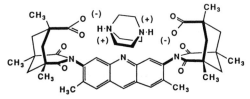

Figure 3   Complex formed with DABCO

　　An unexpected dividend of the structure of the acridine-
derived diacid developed from its acid/base chemistry in polar
solvents such as methanol. Proton transfer occurs within the
cleft and the interior of the molecule becomes a highly polar
microenvironment. At the same time, the outer surface of the
molecule is lipophilic as would be expected for a compound coated
with so many methyl and methylene groups. The large, flat aromatic surface of the acridine nucleus is yet a third domain in this
structure. The juxtaposition of these very different domains
results in the peculiar behavior of this molecule. For example,
the acridine diacid **selectively extracts certain zwitterionic
amino acids into** CHCl3.[4] Moreover, there is high selectivity for
the complexation of β-phenethyl amines such as dopamine, isoquinolines and tryptamine.[5] The spacing between the aromatic surface
and the ammonium head group in these structures is complementary
to that of the acridine surface, and both polar and aromatic
stacking interactions can stabilize the complexes. We have used
this special affinity to achieve the selective transport of aromatic amino acids across simple organic liquid membranes.

　　A special reactivity of the acridine diacid can be observed
with hemiacetals that fit within its cleft. For example, the
dimer of glycol aldehyde is rapidly dissociated in the presence
of catalytic amounts of the diacid, i.e., true turnover occurs.[6]
A likely sequence of events involves the formation of a complex
followed by concerted acid base chemistry either in conventional
sense (in which acid and base approach the substrate in parallel
directions) or in alternate senses. These would involve the acid
and base functions in perpendicular or antiperiplanar arrangements
(Fig. 4). The ability of this acridine diacid to expose both
acids and bases on its surface, yet prevent their collapse upon
one another may be the key to its success as a catalyst. Perhaps
this is also an important feature of enzyme active sites: a
constellation of functionality poised for reaction but kept from
direct interaction by the scaffolding.

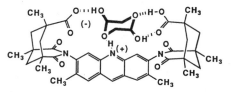

<u>Figure 4</u>　　Proposed concerted catalysis of hemiacetal cleavage.

　　We are pursuing this notion with suitable spacers and in a
number of catalytic contexts. One of these involves developing
enolization catalysts for ketones. Progress in this regard has
been made using the intramolecular system shown. It is rapidly
assembled from the Kemp triacid and amino ketones; the latter are

available from the Dakin-West reaction of α-amino acids (Fig. 5).
Ionization of the carboxylic acids leads to systems in which the
more basic **syn** lone pair is directed toward the enolizable
α-hydrogens.[7]

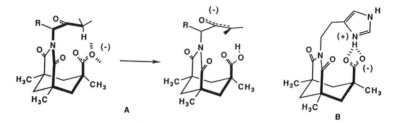

Figure 5   A) Intramolecular general base catalysis of enolization.
           B) Imidazolium-carboxylate pair

     The enhanced enolization rates shown by this system provide
support for Gandour's hypothesis that the efficient use of the
more basic lone pairs is a key feature of enzyme active sites.   A
related system, involving the histamine derivative (Fig. 5B),
reveals the modification of both the imidazole and carboxylate
functions through their mutual contact.[8]   This is also in accord
with Gandour's predictions and is supported by recent mutagenesis
experiments[9] on the serine proteases.

     Lactonization reactions, again involving the more basic **syn**
lone pair, have has also been studied.   The successful application
of these systems to problems posed by stereoelectronic effects at
carboxyl oxygen can be traced to the Kemp triacid and its unique
ability to direct carboxyl functions toward other parts of
molecules derived from it.

     Use of a smaller spacer group, such as meta xylidine diamine,
leads to carboxylic acid chelating agents which show extraordinary
affinity for alkaline earth ions.[10]   Again, the **syn** lone pairs can
be brought to bear on the metal ion (Fig. 6) in contrast to the
many other agents such as EDTA in which only the **anti** lone pairs
are involved.   The **trans** arrangement of the ligands on the metal
in the new complexes is expected to result in some unusual
properties of the ions.

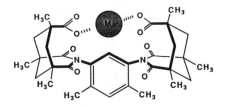

Figure 6   New metal chelate complex with **trans** ligands

A variety of flexible spacer elements is available from
α,ω-diamino-alkanes.  These lead to dicarboxylic acids which
exhibit some promiscuity toward diamines as complexation partners.
A large, rigid spacer represented by the porphyrin nucleus has
also been prepared, through the efforts of J. Lindsey at
Carnegie-Mellon University.  In this structure up to four
carboxylic acids can converge (Fig. 7).  The distances between
acid sites are appropriate for the binding of 4,4'-bipyridine, and
2:1 complexes are observed with this guest species.  One
bipyridine is held above the surface of the porphyrin and another
is held perpendicular to it beneath the surface.[11]

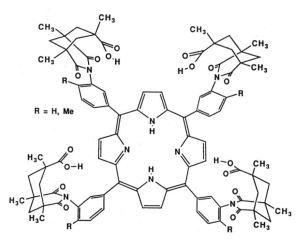

Figure 7   Tetraaryl porphyrin with four convergent carboxyls

The advantage of convergent functional groups and their moderately directional hydrogen bonding characteristics can be exploited in binding other types of substrates. Nucleic acid components can be complexed within clefts having the appropriate lining and shape. For example, the imide of Kemp's triacid can be converted to a number of aromatic derivatives, each of which provide a hydrogen bonding edge that resembles that of thymine and an aryl surface for stacking interactions. Such structures feature binding forces which **converge from perpendicular directions** and they provide unique microenvironments for adenine derivatives.[12] Even molecular chelation of adenine can be arranged through simultaneous Watson–Crick, Hoogsteen and bifurcated hydrogen bonding arrays (Fig. 8A). The affinity is so high that adenosine can be extracted from water and transported across simple liquid membranes with these synthetic molecules as carriers.[13] By altering the hydrogen bonding properties through reduction with $NaBH_4$, preference for binding cytosines (Fig. 8B) can be engineered into the hosts.[14] The weak, intermolecular forces that act over short distances are the key to the high selectivity shown by these model receptors and, quite likely, by the natural ones which they imitate.

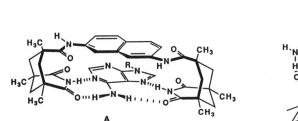

Figure 8   A) Molecular chelation of adenine derivatives
           B) Complexation of cytosines

The positioning of nucleic acid components on these molecular surfaces can be accomplished with a high degree of precision by further tuning through remote steric effects. We are now using this system as a template for chemical reactions, and it is our intent to develop systems capable of self–replication and autocatalytic behavior. For example, binding of adenine derivatives in the manner shown results in intramolecular acyl transfer followed by cis/trans isomerization of the amide (Fig. 9). The resulting product acts as a template for the reaction leading to its own formation; i.e., self–replication occurs. In the presence of diacylamino pyridine the competition for the binding surfaces inhibits the acyl transfer reactions.[15]

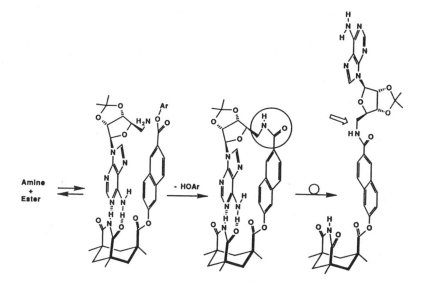

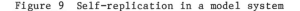

Figure 9   Self-replication in a model system

Finally, we are applying the control of 3-dimensional space
to some problems of asymmetric synthesis.  For example, the
extended aromatic shelf in derivatives such as shown (Fig. 10)
provides an impenetrable barrier to the approach of reagents from
the bottom of the structure.  Accordingly, high enantioselectivity
is achieved in alkylation reactions of enolates.[16]  Cycloaddition
reactions of related derivatives have shown equally impressive
results in the laboratories of Professor Curran[17], and results
with asymmetric protonation augur well for the use of these
principles as general solutions to problems of selectivity.

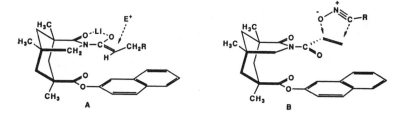

Figure 10   Alkylation (A) and cycloaddition(B) reactions with new
            chiral auxiliaries

In conclusion, it appears that the advantages offered by
cleft-like shapes have justified our abandonment of macrocyclic
compounds. The next generation of structures is likely to
incorporate clefts in which recognition and catalysis merge.

## ACKNOWLEDGEMENT

I am grateful for the valuable experimental assistance of my
coworkers. Their names appear on the original publications.
Financial support was provided by the National Institutes of
Health, the National Science Foundation and Year Laboratories.

## REFERENCES

1. G. Trainor and R. Breslow, J. Am. Chem. Soc., 1981, 103, 154;
   V. T. D'Souza and M. L. Bender, Acc. Chem. Res., 1987, 20,
   146; J.-M. Lehn, Science, 1985, 227, 846; D. J. Cram, ibid.,
   1983, 219, 1177.
2. D. S. Kemp and K. S. Petrakis, J. Org. Chem., 1981, 46, 5140.
3. J. Rebek, Jr., B. Askew, M. Killoran, D. Nemeth and F.-T. Lin,
   J. Am. Chem. Soc., 1987, 2426.
4. J. Rebek, Jr., B. Askew, D. Nemeth and K. Parris, ibid., 1987,
   109, 2432.
5. J. Rebek, Jr., B. Askew, P. Ballester and A. Costero, ibid.,
   1988, 110, 923.
6. J. Wolfe, D. Nemeth, A. Costero and J. Rebek, Jr. J. Am. Chem.
   Soc., 1988, 110, 3327.
7. B. M. Tadayoni, K. Parris and J. Rebek, Jr., J. Am. Chem. Soc.,
   1988, 110, 6575.
8. J. Huff, B. Askew, R. J. Duff and J. Rebek, Jr., J. Am. Chem.
   Soc., 1988, 110, 5908.
9. C. S. Craik, S. Roczniak, C. Largman and W. J. Rutter, Science,
   1987, 237, 909.
10. L. Marshall, K. Parris, J. Rebek, Jr., S. V. Luis and M. I.
    Burguete, J. Am. Chem. Soc., 1988, 110, 5192.
11. J.S. Lindsey, P. C. Kearney, R. J. Duff, T. P. Tjivikua and J.
    Rebek, Jr., J. Am. Chem. Soc., 1988, 110, 6575.
12. J. Rebek, Jr., B. Askew, P. Ballester, C. Buhr, S. Jones, D.
    Nemeth and K. Williams, ibid., 1987, 109, 5033. For another
    system using macrocyclic structures for nucleic acid recogni-
    tion see A. D. Hamilton, D. Van Engen, ibid., 1987, 109, 5035.
13. T. Benzing, T. Tjivikua, J. Wolfe and J. Rebek, Jr., Science,
    1988, 242, 266.
14. K.-S. Jeong and J. Rebek, Jr. J. Am. Chem. Soc., 1988, 110,
    3327.
15. T. P. Tjivikua, unpublished observations.
16. K. S. Jeong, K. Parris, P. Ballester and J. Rebek, Jr.
    submitted for publication.
17. D. Curran and T. Heffner, unpublished observations.

# The Basis for Molecular Recognition

## By P. Kollman

DEPARTMENT OF CHEMISTRY AND PHARMACEUTICAL CHEMISTRY, UNIVERSITY OF CALIFORNIA, SAN FRANCISCO, CA 94143, USA

## Background

What do we mean by molecular recognition? Molecular recognition refers to the fact that a receptor molecule binds some molecules more tightly than others, due to the fact that the non-covalent interactions between the receptor and some molecules are more favorable.

What is the basis for molecular recognition? Fersht has pointed out in this symposium that there are four types of interactions that contribute: electrostatic (ionic), hydrogen bonding, van der Waals and hydrophobic. [1] We agree that these all play a role; it is the purpose of this article to describe the theoretical underpinning for these interactions and to describe, with examples, how they play a role in determining non-covalent interactions between small molecules and ionophores and between proteins and ligands.

Although the basis for molecular electronic structure and properties is inherently quantum mechanical, there is increasing recognition of the fact that one can simulate rather accurately molecular geometries, conformational energies and non-covalent interactions with rather simple analytical potential functions such as (1): [2, 3]

$$E_{total} = \sum_{bonds} K_r \, (r - r_{eq})^2 + \sum_{angles} K_\theta \, (\theta - \theta_{eq})^2 + \sum_{dihedrals} \frac{V_n}{2}[1 + \cos(n\phi - \gamma)] + \tag{1}$$

$$\sum_{i<j} [\frac{A_{ij}}{R_{ij}^{12}} - \frac{B_{ij}}{R_{ij}^6} + \frac{q_i q_j}{\varepsilon R_{ij}}] + \sum_{H-bonds} [\frac{C_{ij}}{R_{ij}^{12}} - \frac{D_{ij}}{R_{ij}^{10}}]$$

Why does such an approach work? One can use more complex functions than (1) to describe highly strained molecules, [4] but in any case, such equations rely on the transferability of bonding properties for given functional groups in organic/biochemical molecules. In other words, the properties of a C-C bond, a C-C-C bond angle, a torsional rotation around a C-C bond and the non bonded properties of an $sp^3C$ atom are similar enough in any organic molecule to allow one to calibrate (parameterize) the properties of these fragments to experiment on a few very well described small molecules and then to successfully use such parameters on a wide variety of other molecules, not part of the calibration set. Thus, many of the parameters in equation (1) are fit to empirical data. However, to derive the partial charges on the atoms $q_i$ require fitting the quantum mechanically derived electrostatic potential to classical charges. [5] Since these charges are critical to the hydrogen bonding and electrostatic part of molecular recognition, making them correspond to reality as closely as possible is critical. Thus, in summary, using theoretical approaches such as equation (1) to study molecular recognition requires careful calibration of the parameters in such an equation to empirical or quantum mechanical data, whichever is most relevant or appropriate. For example, the most successful models of liquid water are empirically derived, varying the charges and van der Waals parameters on oxygen, with compensating charges on hydrogen, in order to reproduce the density and enthalpy of vaporization of the liquid. [6]

Once one has a set of parameters for equation (1), how can one use these to study molecular recognition? The thermodynamic cycle (2) provides a method to analyze molecular recognition, as we defined it in the first paragraph:

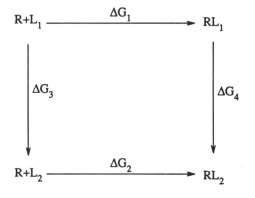

$$\Delta\Delta G_{bind} = \Delta G_2 - \Delta G_1 = \Delta G_4 - \Delta G_3 \tag{3}$$

The free energy for the receptor molecule to bind ligand $L_1$ is $\Delta G_1$ and for it to bind

ligand $L_2$ is $\Delta G_2$. The relative free energy of binding $\Delta\Delta G_{bind}$ is the difference between

these free energies. The larger $\Delta\Delta G_{bind}$, the greater the molecular recognition of the

receptor for $L_1$ compared to $L_2$ (the more negative $\Delta G_i$, the tighter the binding of $L_i$ to

R). It is usually quite difficult to calculate $\Delta G_1$ and $\Delta G_2$ by computer simulation

methods, because the processes usually involve large and complicated structural/solvent

changes and the breaking and forming of many interactions. Thus, one uses the fact that

$\Delta G$ is a state function, i.e. the sum of the free energies around the cycle is zero and evalu-

ates instead $\Delta G_4$ and $\Delta G_3$. $\Delta G_4$ corresponds to "mutating" $L_1$ into $L_2$ when they are

bound to the receptor R and $\Delta G_3$ corresponds to the mutation when $L_1$ (and $L_2$) are free

in solution.

How does one carry out this chemical "alchemy" of mutation? One gradually

changes the molecular mechanical parameters for $L_1$ into those for $L_2$, using Monte Carlo or molecular dynamics to generate an ensemble average of the system (eqn (4))

$$\Delta G \ (A \rightarrow B) = -RT \ \ln \ < e^{-\Delta H/RT} >_A \qquad (4)$$

The free energy for mutating system A into B can be calculated by generating a representative set of configurations of A, and ensemble average $< >_A$ and averaging $e^{-\Delta H/RT}$ over this ensemble, where $\Delta H$ is the difference in Hamiltonia between the systems $\Delta H = H_B - H_A$, R the gas constant and T the absolute temperature. [7] A critical paper in this regard was the demonstration by Ravimohan and Jorgensen using Monte Carlo [8] that one could calculate the relative free energy of solvation of methanol and ethane ($\Delta G_3$) in excellent agreement with experiment. In practice, if A and B differ significantly, as did methanol and ethane in their solvation properties, one creates a number of fictitious systems that are part way between A and B and evaluates the free energies for each of these transformations. The total free energy change $\Delta G \ (A \rightarrow B)$ then becomes eqn (5).

$$\Delta G = \sum_\lambda = \Delta G_\lambda \qquad (5)$$

where $\lambda = 0$ corresponds to system A and $\lambda = 1$ corresponds to system B. These methodologies [7,9] are more fully described in ref. 7 and 9.

When one analyzes molecular recognition using a cycle like (2), it becomes immediately apparent that molecular recognition must be considered a function of the solvent. The ability of a receptor R to discriminate between two ligands $L_1$ and $L_2$ depends not only on the relative free energies of association of these ligands to the receptor ($\Delta G_4$), but also on the relative solvation free energies of these ligands, $\Delta G_3$. The latter can be understood by realizing that, the higher the free energy cost of desolvating a ligand, the weaker its association with the receptor. Although this qualitative statement is obvious, a major development in theoretical chemistry of complex molecules has been the capability to calculate both $\Delta\Delta G_{bind}$ and $\Delta G_3$ in good agreement with experiment, and thus enable one to derive mechanistic insights into these processes. In fact, we have argued that the basis for molecular solvation (and, by implication, ligand-receptor non-covalent interactions) can be quantitatively and qualitatively understood using the terms in equation (1): the $q_i q_j$ term can rationalize ionic electrostatic and hydrogen bonding recognition (the 10-12 H bond term serves mainly to fine tune H-bond distances and con-tributes little to H-bond attraction), the $B_{ij}$ term (dispersion attraction) contributes to specificity through van der Waals/dispersion attraction and, finally, the $A_{ij}$ term contri-butes to specificity through exchange repulsion/size effects as well as providing the basis for the hydrophobic effect by contributing to the solvation free energy $\Delta G_3$.

Thus, of the four interactions mentioned by Fersht, the first three (ionic, H-bonding and van der Waals) come directly from the interaction terms in equation (1), whereas the hydrophobic interaction comes mainly from a favorable <u>desolvation</u> effect of non-polar residues in water solvent. But all of them can be quantitatively calculated with equations (1), (4) and (5), provided solvent is explicitly included in the calculations.

Applications

We now present some applications of the free energy approach to problems in molecular recognition, with examples of small organic receptors as well as protein recep-tors and various ligands including ions, small organic molecules, and diatomic gases and protein inhibitors.

Grootenhuis and Kollman (GK) [10] have used free energy approaches to study the interaction of dibenzo 18-crown-6 (PB186) and dibenzo-30-crown-10 (DB3010) with various metal cations $Na^+$, $K^+$, $Rb^+$ and $Cs^+$ in methanol. The relative experimental affinities are, for DB186, a preference for $K^+$ over $Na^+$ and $Rb^+$ by approximately equal amounts, and for DB3010, approximately equal $K^+$ and $Rb^+$ affinities, with $Cs^+$ less tightly bound and $Na^+$ much less tightly bound. GK used cycle (2) to study these systems with R=DB186 or DB3010 and $L_i$ = $Na^+$, $K^+$, $Rb^+$ and $Cs^+$. As has been noted before for internal energies $\Delta E$, both $\Delta G_3$ and $\Delta G_4$ are positive when one increases the size of the cation, so $\Delta\Delta G_{bind}$ is a delicate balance between the solvation effects ($\Delta G_3$) and the binding effects ($\Delta G_4$). The calculations are able to reproduce all the main experimental trends, in that for R = DB186, $K^+$ is calculated to be most tightly bound, with $Na^+$ and $Rb^+$ comparably less bound and $Cs^+$ least tightly bound. For R = DB3010, $\Delta\Delta G_{bind}$ for $L_1 = K^+$ and $L_2 = Rb^+$ is close to zero, whereas it is very large for $L_1 = K^+$ and $L_2 = Na^+$ and significant when $L_1 = Rb^+$ and $L_2 = Cs^+$. The $\Delta\Delta G_{bind}$ calculated are far from quantitative, due to a combination of having less than perfect potential functions (1) and the use of a very simple solvation model, in which a small cluster of six methanol molecules was used in the calculation of $\Delta G_3$ and two methanol molecules in the calculation of $\Delta G_4$. Larger clusters did not lead to quantitative agreement with experiment, so potential function inaccuracy or the need to use periodic boundary conditions are the most likely reasons for the discrepancies. Nonetheless, all the chemical trends in the calculated $\Delta\Delta G$ are reproduced.

Another application of cycle (2) to crown interactions was the analysis of the binding of malonitrile, acetonitrile and mitromethane to 18-crown-6 (186) in benzene solution [11]. Here, one was able to approximate $\Delta\Delta G_{bind} = \Delta G_4$ and not use any benzene molecules in the calculation of this quantity. This was because benzene interacts so weakly with both R (186) and $L_1$ - $L_3$. In fact, we used normal mode analysis to calculate $\Delta G_1$ directly for $L_1$ = each of the three ligands and found values of $\Delta G_1$ that were in reasonable agreement with experiment. The fact that one can get such agreement further validates the potential functions used and the fact that the solvent (benzene) interactions with the solutes are so weak. It also demonstrates that one can calculate absolute $\Delta G$ for molecular association in favorable cases in good agreement with experiment.

Lopez [12] has used cycles (6) and (7) to study the relative binding affinities of $O_2$ and CO to model porphyrins.

(6)

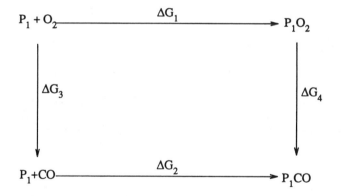

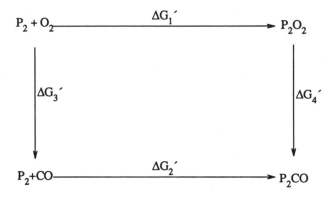

The relative binding affinity of $O_2$ and $CO$ to prophyrins $P_1$ and $P_2$ are given by

$$\Delta\Delta G_{bind} = \Delta\Delta G_2 - \Delta G_1 = \Delta G_4 - \Delta G_3 \qquad (8)$$

$$\Delta\Delta G'_{bind} = \Delta\Delta G'_2 - \Delta G'_1 = \Delta G'_4 - \Delta G'_3 \qquad (9)$$

$$\Delta\Delta G = \Delta\Delta G'_{bind} - \Delta\Delta G_{bind} = \Delta G'_2 - \Delta G'_1 - (\Delta G_2 - \Delta G_1) \qquad (10)$$

$$= \Delta G'_4 - \Delta G_4 \; (because \, \Delta G_3 = \Delta G'_3)$$

If $P_1$ is a reference porphyrin with no "strap", then one can evaluate the effect of any organic "strap" attached to the porphyrin by calculating $\Delta G'_4$ and $\Delta G_4$ in which one mutates the charge distribution of $FeO_2 \rightarrow FeCO$. Such an approach cannot calculate the relative bond energies of Fe to CO or $O_2$ (these are presumed to cancel in cycles (6) and (7), but the effect of the surrounding "straps" on the relative binding energies. In two of these cases, the calculated $\Delta\Delta G$ is in good agreement with experiment. In the third case, the agreement is not so good, and we have suggested that this could be a solvent effect, the calculations having been done in the gas phase and the experiments in benzene solvent. Since some of the straps contain CONH groups and the NH parts of these could be interacting either with benzene or the $O_2$ ligand; (which has a large negative charge than the CO) the neglect of solvent in the calculation would lead to an overestimate of the binding stabilization of $O_2$, as found. Further application of this approach to myoglobin (a protein strap) and its site specific mutants is in progress.

Applying cycle (2) to inhibitors of thermolysin of formula CBZ - Gly$^P$ - X - Leu - Leu, with CBZ ≡ carbobenzoxy, Gly$^P$ ≡ Glycine with $PO_2^-$ replacing the sciscile bond and X = NH,O or $CH_2$, has proven very fruitful. In earlier calculations [13], with R = thermolysin and $L_1$ = the above ligand with X=NH and $L_2$ the ligand with X=O, $\Delta\Delta G_{bind}$ experimentally was found to be 4.1 ± 0.1 [14] and the calculated value for this quantity was 4.2 ± 0.5. Furthermore, X-ray structure suggested that the two inhibitors were binding essentially identically in the enzyme site. The X-ray structures [15] suggested that the Ala 113 C = O functionality was in a position to H-bond to the the X=NH group, whereas an X=O group would be unable to form this H-bond. This led Bartlett [14] to interpret this 4 kcal/mole as an intrinsic H-bond energy of a protein NH ... OC interaction. Our calculations and the X-ray structure suggested a somewhat different interpretation. $\Delta G_4$ was calculated to be 7 kcal/mole and interpreted as due to both an NH ... OC attraction and an O ... OC "forced repulsion". This repulsion was "forced" because the favorable $PO_2^-$ and two leucines anchor the X=O oxygen of the ligand in an unfavorable place near the Ala 113 carbonyl oxygen. Furthermore, Bash *et al.* [13] calculated a $\Delta G_3$ = 3 kcal/mole, suggesting the relative solvation effects could be important. To test these interpretations/hypotheses, we carried out simulations on X=$CH_2$, agreeing with Paul Bartlett that we would complete the calculations and submit a paper on these prior to syntheses/testing of such analogs. Thus, we would test the ability of the theory to be truly predictive. If the Bartlett interpretation were valid, the X=$CH_2$ and X=O compounds should bind similarly, since neither can form an H-bond. On the other hand, if solvation and "forced repulsion" effects are important, the X=CH$_2$ compound should bind more tightly than X=O. Interestingly, the calculations found that for L$_1$, X=NH and L$_2$, X=CH$_2$, that $\Delta G_4$ = 2 kcal/mole and $\Delta G_3$ ~ 2 kcal/mole. This supports the importance of desolvation and forced repulsion effects. Thus $\Delta\Delta G_{bind}$ for NH vs CH$_2$ was calculated to be 0.0-0.3 kcal/mole and approximately 4 kcal/mole for CH$_2$ vs O, [16] in excellent agreement with subsequent experiments by Bartlett and co-workers. [17] This has been an important validation of the free energy method in studies of biological macro-molecules.

Studies on relative solvation effects on organic model systems have been insightful. In such studies, one uses thermodynamic cycle (11).

(11)

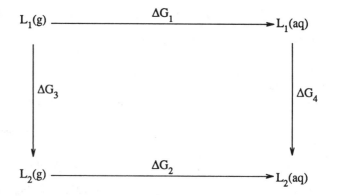

$$\Delta\Delta G_{solv} = \Delta G_2 - \Delta G_1 = \Delta G_4 - \Delta G_3 \qquad (12)$$

In practice, one can assume intramolecular effects are zero or identical in gas and solution and approximate $\Delta\Delta G_{solv} = \Delta G_4$, where only solute-solvent energies contribute to

$\Delta G_4$.

If $L_1$ = nothing and $L_2 = CH_4$, the calculated $\Delta \Delta G_{solv}$ = 2 kcal/mole, in agreement with experiment. [18] This validates that molecular mechanical energy functions have all the necessary information to calculate "hydrophobic" effects, provided that solvent $H_2O$ is included in the calculations. The fact that $\Delta \Delta G_{solv}$ is unfavorable (positive) comes from the exchange repulsion ($A_{ij}/R_{ij}^{12}$) term in equation (1). A physical picture for this is that water molecules in the vicinity of the "growing" methane molecule will experience exchange repulsion in the presence of the methane, because in doing so, they can improve the water-water H-bonding near the methane. If $L_1 = CH_4$ and $L_2 = C_2H_6$, $\Delta \Delta G_{solv}$ = -0.2 kcal/mole[19] and if $L_1 = C_2H_6$ and $L_2 = C_3H_8$, $\Delta \Delta G_{solv}$ = 0.2 kcal/mole. We argue [18] that these small numbers are due to the balance between exchange repulsion ($\dfrac{1}{R^{12}}$) and dispersion ($\dfrac{1}{R^6}$) terms. In the case of $L_1$ = dummy and $L_2 = CH_4$, the exchange repulsion dominates because a large group is being created in the solvent. In the case of $L_i$ = methane and $L_2$ = ethane or $L_1$ = ethane and $L_2$ = propane, $\Delta \Delta G_{solv}$ is very small because both dispersion and exchange repulsion terms are of comparable magnitude. In non-aqueous solvents, one expects that the repulsion term is going to be significantly smaller than the dispersion term and all these $\Delta \Delta G_{solv}$ values will become more negative.

The relative aqueous solvation contributions for adding a $CH_3$ group to a molecule are not always unfavorable, as we see for going from $L_1 = CH_4$ to $L_2 = C_2H_6$. A second example comes from the three molecules, acetamide, N-Methyl acetamide, and N,N dimethyl acetamide. The solvation free energy becomes more negative upon adding the first methyl group and then positive upon adding the second, both in simulations [18] and experiments. All three molecules have a similar dipole moment. Again, the balance between exchange repulsion and dispersion is very delicate in these systems.

Another example of "non-additivity" in solvation free energies comes from considering the solvation free energy of benzene, nitrobenzene, phenol and p-NO$_2$ phenol. p-NO$_2$ phenol is 5 times more soluble than one would predict based on the solubility of NO$_2$-benzene and phenol compared to benzene. [18] One can simulate this quantitatively using free energy perturbation approaches by constructing a model charge distribution of nitrophenol from nitrobenzene and phenol without allowing any resonance effect and also a correct charge distribution allowing the resonance effect and then mutating one charge distribution into the other. The calculated $\Delta\Delta G$ for this mutation, -0.9 kcal/mole, is in reasonable agreement with this non-additive solvation effect.

## Summary

We have applied free energy perturbation approaches to a wide variety of other systems, e.g. enzyme catalysis, [20] protein stability, [21] DNA sequence dependent stabilities [22] and DNA-drug binding. [23] It has proven to be useful in these cases as well. It is clear that molecular recognition can be correctly analyzed by including solvent effects explicitly and using modern simulation techniques. For many non-covalent interactions, only electrostatic, exchange repulsion and dispersion terms need to be included to accurately represent such recognition. Of course, many highly charged systems and those involving transition metals require a more complex and sophisticated interaction potential, the development of which is an area of current research. Nonetheless, the prospects for the use of computer simulation techniques to give useful numbers and insights is great.

Acknowledgements

We acknowledge research support from the NIH (GM-29072), DARPA (under contract N00014-86-K-0757 administered by the Office of Naval Research, R. Langridge, P.I.) and the use of the facilities of the UCSF Computer Graphics Laboratory, R. Langridge, director, supported by RR-1081 from NIH. PAK is pleased to acknowledge his collaborators, whose work is described in the references.

1.  A. Fersht, (Lecture at this symposium).

2.  S. J. Weiner, P. A. Kollman, D. A. Case, U. C. Singh, C. Ghio, G. Alagona, S. Profeta Jr and P. Weiner, *J Amer Chem Soc* **106**, 765-784 (1984).

3.  S. J. Weiner, P. A. Kollman, D. T. Nguyen and D. A. Case, *J Comp Chem* **7**, 230-252 (1986).

4.  N. L. Allinger, *J Amer Chem Soc*, pp. 8127-8134 (1977).

5.  U. C. Singh and P. Kollman, *J Comp Chem* **5**, 129 (1984).

6.  W. Jorgensen, J. Chandrasekhar, J. Madura, R. Impey and M. Klein, *J Chem Phys* **79**, 926-935 (1983).

7.  U. C. Singh, F. K. Brown, P. A. Bash and P. A. Kollman, *J Amer Chem Soc* **109**, 1607 (1987).

8.  W. Jorgensen and C. Ravimohan, *J Chem Phys* **83**, 3050 (1985).

9.  D. Beveridge and F. DiCapua, Free Energy via Molecular Simulation: Applications to Chemical and Biomolecular Systems, *Ann Rev Biophys Bioengin*, D. Engelman

Ed. (1989).

10. P. D. J. Grootenhuis and P. A. Kollman, *J Amer Chem Soc* **111**, 2152 (1989).

11. P. D. J. Grootenhuis and P. Kollman, *J Amer Chem Soc* **111**, 4046 (1989).

12. M. Lopez and P. Kollman, Application of Molecular Dynamics and Free Energy Perturbation Methods to Metalloporphyrin Ligand Systems. I. CO and Di-oxygen Binding to Four Heme Systems, *J Amer Chem Soc* (in press).

13. P. Bash, U. C. Singh, F. Brown, R. Langridge and P. Kollman, *Science* **235**, 574 (1987).

14. P. A. Bartlett and C. K. Marlowe, *Science* **235**, 569 (1987).

15. D. E. Tronrud, H. M. Holden and B. W. Matthews, *Science* **235**, 571 (1987).

16. K. Merz and P. Kollman, Free Energy Perturbation Simulations of the Inhibition of Thermolysin, *J Amer Chem Soc* (in press).

17. P. Bartlett and B. P. Morgan, (unpublished, see note added in proof in ref. 16.).

18. P. Bash, U. C. Singh, R. Langridge and P. Kollman, *Science* **236**, 564 (1987).

19. Y. Marcus and A. Ben Naim, *J Chem Phys* **81**, 2106 (1984).

20. S. Rao, P. Bash, U. C. Singh and P. Kollman, *Nature* **328**, 551-554 (1987).

21. L. Dang, K. Merz, Jr and P. Kollman, Free Energy Calculations of Protein Stability: The Thr 157 → Val 157 Mutation to T4 Lyzozyme, *J Amer Chem Soc* (in press).

22. D. Pearlman and P. Kollman, The Calculated Free Energy Effects of 5 Methyl Cytosine on the B to Z Transition in DNA, *Biopolymers* (in press).

23. P. Cieplak, S. Rao, P. Grootenhuis and P. Kollman, Free Energy Calculations on Base Specificity of Daunomycin and Acridine-DNA Interactions, *Biopolymers* (in press).

# Design of Artificial Receptors for Biochemically Interesting Substrates

By Andrew D. Hamilton, Suk-Kyu Chang, Shyamaprosad Goswami, Alex V. Muehldorf, and Donna Van Engen

DEPARTMENT OF CHEMISTRY, UNIVERSITY OF PITTSBURGH, PITTSBURGH, PA 15260, USA

The design and synthesis of artifical receptors able to form strong and selective complexes to small organic substrates is an area of intense current interest. Crucial to the success of this endeavor is the ability to incorporate directed binding interactions (hydrogen bonding, π-stacking, electrostatic etc.) within a cleft or cavity of complementary size and shape to the substrate.

## Directed Hydrogen Bonding Interactions

A crucial binding interaction in biological recognition involves hydrogen bonding between the protein and its substrate. A synthetic cavity containing several directed hydrogen bonding groups should lead to binding and potential orientation of a substrate with complementary groups. Our first receptors based on this strategy involved as substrates the barbiturate family of drugs (e.g. 1). These are attractive targets for molecular recognition studies due to their widespread use as sedatives and anticonvulsants and their simple and rigid arrangement of hydrogen bonding groups. All six of the accessible hydrogen bonding sites (four CO lone pairs and two imide NHs) can be complexed by two 2,6-diamino-pyridine units incorporated into a macrocyclic cavity of appropriate size (fig. 1). An important design question con-

1a. R$_1$=R$_2$=Et,R$_3$=H
 b. R$_1$=Et,R$_2$=Ph,R$_3$=H
 c. R$_1$=Et,R$_2$=Ph,R$_3$=CH$_3$
 (Racemic)

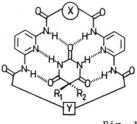

Fig. 1

cerns not only the positioning of the hydrogen bonding groups but
also the rigidity of the supporting framework.  If the receptor is
too flexible then intramolecular hydrogen bonds may occur and the
cavity may collapse.

The first barbiturate receptors (e.g. 2 and 3) employed on
isophthalate spacer group and were prepared in a simple two-step
sequence from 2,6-diaminopyridine.[1]  The nature of the binding
cavity in 3 was confirmed by X-ray crystallography which shows
(fig 2) an open conformation for the macrocycle with all six hydrogen bonding groups directed toward its center.  The receptor is not

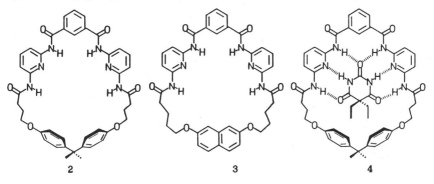

|     |     |     |
|-----|-----|-----|
| **2** | **3** | **4** |

fully preorganized (i.e. planar) but has a twisted conformation
with a 44° angle between the pyridine rings.[2]  In addition, a THF
molecule of crystallization is occupying the cavity and forms a
single hydrogen bond to an amide-NH.  The binding properties of 2
and 3 can be readily followed by [1]H NMR.  Addition of one equivalent
of barbital 1a to a $CDCl_3$ solution of 2 or 3 results in large downfield shifts of the host amide and guest imide proton resonances
(1.65, 1.63 and 4.38 ppm, respectively for 2).  Such large shifts
are strongly indicative of a hexahydrogen bonded complex of the
type shown in 4.  Continuing the addition of 1 into 2 or 3 leads to
a titration curve that shows clear saturation with a sharp break at
a guest:host ratio of 1:1.  Further analysis of the binding curve
using either a Scatchard or non-linear regression approach[3] yields
association constants ($K_a$) for the interaction which are collected
in table 1.  These show that for a complementary substrate such as
barbital large $K_a$ values (1.37 x $10^6$ $M^{-1}$  and 1.35 x $10^5$ $M^{-1}$) are
possible.  Enforced removal of 2-3 hydrogen bonds from the interaction, as with mephobarbital, results in a $10^3$  fold drop in binding.
Further information on the recognition process has been obtained by
X-ray crystallography.  The structure of the complex between 1a and
3 confirms the hexahydrogen bonded nature of the interaction between receptor and substrate (fig. 3).  The bond lengths of the six
hydrogen bonds fall into three pairs corresponding to short (~2.9Å;
N(4)-O(41), N(3)-O(43)), medium (~3.0Å; N(5)-N(7), N(2)-N(8) and

236                                    *Molecular Recognition*

**Table 1:** Association Constants for the Receptor-Barbiturate Interaction

| Receptor | Barbiturate | $K_a, M^{-1}$ (25°C, CDCl$_3$) |
| --- | --- | --- |
| 2 | Mephobarbital 1c | 6.80 x 10$^2$ |
| 3 | Phenobarbital 1b | 2.80 x 10$^5$ |
| 2 | Phenobarbital 1b | 1.97 x 10$^5$ |
| 3 | Barbital 1a | 1.35 x 10$^5$ |
| 2 | Barbital 1a | 1.37 x 10$^6$ |

Fig. 2                                    Fig. 3

long (~3.2Å; N(6,1)-O(42)) hydrogen bonds.[4] The macrocycle has also undergone a conformational change (compared to fig. 2) to bring all six hydrogen bonding groups into the same plane. The barbiturate itself is not coplanar to the macrocycle but lies at a 27° angle relative to the pyridine rings. The skewed position of the guest within the cavity may be due to unfavorable steric interactions between the barbital-2-carbonyl group and the isophthaloyl-2-proton. Non-planar hydrogen bonded complexes of this type are

common in nuclic acid chemistry where propeller twists between
interacting rings of up to 30° are often encountered.[5]

## Directed π-Stacking Interactions

The simultaneous influence of both hydrogen bonding <u>and</u> π-
stacking interactions offers a powerful approach for the recognition of planar heterocyclic substrates such as the nucleotide
bases. By exploiting this <u>two-site</u> binding strategy, we have developed a series of receptors for the nucleotide bases that involves
the perpendicular convergence of hydrogen bonding and aromatic
stacking interactions. Our strategy was to link within a macrocyclic framework, a group capable of stacking with the nucleotide
base to one complementary to its hydrogen bonding periphery (fig. 4).
The geometry of the π-stacking groups is of particular interest
since in a survey of protein crystal structures Petsko[7] has identified two important orientations for aromatic-aromatic interactions;
namely face-to-face and edge-to-face (fig. 5). We have used the
synthetic receptor approach to investigate the importance of the
electronic characteristics of the stacking group on its orientation
in the complex.

### AROMATIC-AROMATIC INTERACTIONS

FACE-TO-FACE          EDGE-TO-FACE

Fig. 4                                    Fig. 5

The first receptors for thymine combined 3,6-disubstituted-
2,7-dialkoxynaphthalenes as π-stacking components with a 2,6-
diaminopyridine unit as hydrogen bonding component (e.g. 5 and 6).[6]
Macrocycle 6, containing two electron withdrawing ester groups on
the naphthalene ring, forms stronger complexes ($K_a$ = 570 $M^{-1}$) than
unsubstituted 5 ($K_a$ = 290 $M^{-1}$) with 1-butylthymine 7. The X-ray
structure of complex 6:7 (fig. 6) shows triple hydrogen bonding
between the thymine and diaminopyridine plus a parallel, face-to-
face interaction between the naphthalene and thymine rings. An
insight into the special stabilization involved in stacking comes
from MNDO calculations on thymine and 2,7-dimethoxynaphthalene-3,6-
dicarboxylate. The resulting charge distributions are superimposed
(sign only) on a downward view of structure 6:7 (fig. 7a). This
shows five points of contact where partially positive charged atoms

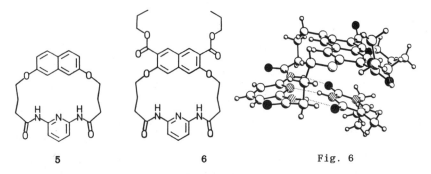

5                            6                         Fig. 6

on the naphthalene precisely align themselves with partially negative regions on the thymine.  Thus, electrostatic interactions
between regions of complementary charge distribution on the rings
play an important role in $\pi$-stacking.  Changing the electronic charge
distribution of the stacking group would then be expected to alter
the orientation.  Replacing the diester groups in 6 by two ether
groups leads to macrocycle 8 which shows substantially weaker binding ($K_a$ = 138 $M^{-1}$) to 7 than does 6.  MNDO calculations on 2,3,6,7-
tetramethoxynaphthalene show a reversal of sign on carbons-4 and -5
leading to a repulsive electrostatic interaction between receptor
and substrate if a face-to-face geometry were to form (fig. 7b).

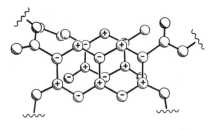

a                            Fig. 7                          b

The X-ray structure of complex 8:7 confirms that the face-to-face
geometry is avoided and that the naphthalene takes up an almost
perpendicular, edge-to-face orientation with respect to the
substrate (fig. 8).  The solution [1]H NMR data on complex 8:7 is
also consistent with this conformation.  Thus, we have shown that
an electrostatic complementarity between partial charges on the
rings can lead to strong face-to-face stacking, while in the
absence of such effects a weaker edge-to-face interaction is preferred.[8]

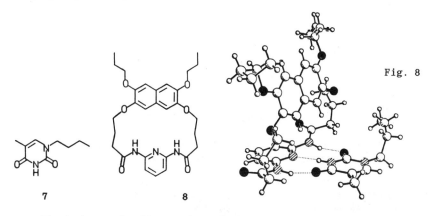

Fig. 8

7          8

The hydrogen bonding region of these receptors can also be varied to give new binding specificities for the other nucleotide bases. 2-Amino-1,8-naphthyridine 9 possesses a triple hydrogen bonding complementarity with the periphery of guanine. We have prepared a macrocyclic receptor 10 containing both an amino-naphthyridine and a π-stacking unit and shown that it binds to guanine derivatives by a combination of aromatic stacking and hydrogen bonding interactions as in 11. The participation of the

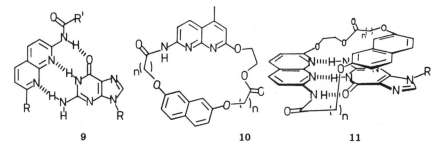

9                    10                    11

π-stacking group is confirmed by an increase in $K_a$ (126 to 502 $M^{-1}$) on going from a simple acyclic aminonaphthyridine to 10.[9] A similar strategy can be applied to the recognition of adenine. Bis-(2-aminopyridine) derivatives 12 can form four hydrogen bonds to the periphery of adenine (involving both Watson–Crick and Hoogsteen interactions). We have prepared a series of macrocyclic receptors 13 combining both a naphthalene π-stacking unit and a 1,2-bis-(2-amino-6-pyridyl)-ethane group and shown that they form strong complexes with 9-alkyladenine 14, (for example, $K_a$ value for 15 (n=3) is 3200 $M^{-1}$).[10] The binding conformation of 15 requires all

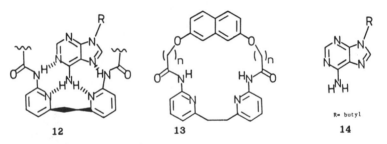

**12**            **13**            **14**

R= butyl

four hydrogen bonding groups of the bis-(2-aminopyridine) group to
be directed towards the center of the cavity. This orientation is
supported by an X-ray structure on uncomplexed **13** (n=2) (fig. 9).
The pyridine-Ns are placed at 5.04Å and the amide-NHs at 7.33Å from
each other in good binding complementarity to the amino group and
purine-Ns of adenine.

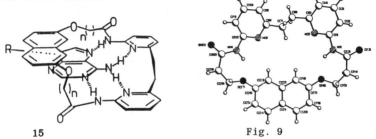

**15**                              Fig. 9

A key step in the recognition of nucleotide bases involves
oligomerization of the individual binding units to form a multiple
receptor for the complexation of oligonucleotide base substrates
(e.g. dimer in fig. 10 compared to monomer in fig. 4). We have
recently prepared a double receptor containing two <u>two-site</u> binding
regions for thymine and shown that it forms strong complexes with
bis-thymine derivatives.

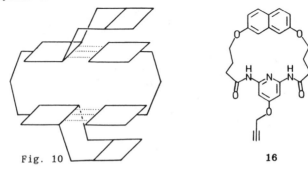

Fig. 10                              **16**

In designing the double receptor, a rigid diyne spacer was positioned between the two binding regions to prevent collapse of the cavity. The synthesis involved an oxidative coupling of propargyloxy-substituted **16** to form a dimeric receptor **17** in 70% yield. In order to study the double binding properties of **19** an organic soluble, bis-thymine derivative **18** containing an analogous diyne spacers was prepared. Double receptor **17** provides a matched fit for bis-thymine **18** in terms of spacer length, hydrogen bonding and aromatic stacking characteristics and should readily form a 1:1 complex of type **19**. This was supported by $^1$H nmr and mass

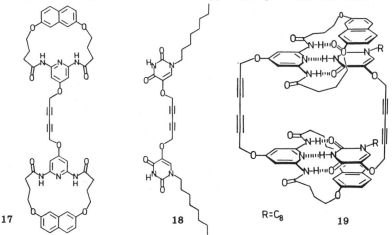

**17**          **18**          R=C$_8$          **19**

spectrometry. Addition of one equivalent of **18** to a CDCl$_3$ solution of **17** gave large downfield shifts of the amide and imide resonances (1.66 and 1.75 ppm) and upfield shifts thymine ring-H and N-CH$_2$ (0.38 and 0.32 ppm) and naphthalene-1,8, -3,6 and -4,5H (0.47, 0.27 and 0.36 ppm), confirming both hydrogen bonding and face-to-face π-stacking interactions in **19**. The titration curve showed 1:1 stoichiometry for the complex and gave an association constant of 2.03 x 10$^4$ M$^{-1}$. Further evidence for the 1:1 character of **19** (rather than a higher homologue) came from fast atom bombardment mass spectrometry. Ionization of a 1:1 mixture of **17** and **18** in a p-nitrobenzyl alcohol matrix gave an M+H$^+$ peak at M/Z = 1472 (fig. 11) with no detectable peaks at higher mass.

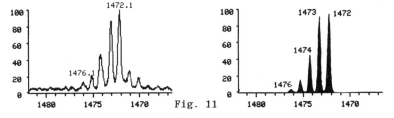

Fig. 11

Acknowledgement.

We thank the National Institutes of Health for their support of the work described in this article.

REFERENCES

1. S. K. Chang and A. D. Hamilton, J. Am. Chem. Soc., 1988, 110, 1318.
2. S. K. Chang, D. Van Engen, and A. D. Hamilton, J. Am. Chem. Soc., in press.
3. R. Foster and C. A. Fife, Prog. Nucl. Magn. Reson. Spectrosc., 1969, 4, 1.
4. R. Taylor, O. Kennard, and W. Versichel, J. Am. Chem. Soc., 1983, 105, 5761.
5. W. Saenger in "Principles of Nucleic Acid Structure", Springer-Verlag, New York, 1984, p. 26.
6. A. D. Hamilton and D. Van Engen, J. Am. Chem. Soc., 1987, 109, 5035.
7. S. K. Burley and G. A. Petsko, Science (Washington, DC), 1985, 229, 23.
8. A. V. Muehldorf, D. Van Engen, J. C. Warner and A. D. Hamilton, J. Am. Chem. Soc., 1988, 110, 6561.
9. A. D. Hamilton and N. Pant, J. Chem. Soc., Chem. Commun., 1988, 765.
10. S. Goswami, D. Van Engen and A. D. Hamilton, J. Am. Chem. Soc., in press.

# Structure-directed Synthesis of Unnatural Products

By Franz H. Kohnke, John P. Mathias, and J. Fraser Stoddart

DEPARTMENT OF CHEMISTRY, THE UNIVERSITY, SHEFFIELD S3 7HF, UK

## 1 INTRODUCTION

Although the molecules of life, and the natural products associated with them, reveal a panoply of molecular structures, it is patently obvious that, once upon a time, certain key, relatively-simple, chemical compounds served[1] spontaneously as the building blocks for the elaboration of apparently complex biomolecules, or their biosynthetic precursors. Against this truly remarkable back-cloth of molecular evolution, which heralded the beginning of life itself, it does not seem too unrealistic to assume that there is a world of new materials composed of essentially unnatural products waiting to be conceived and constructed by chemical scientists.

## 2 STRUCTURE-DIRECTED SYNTHESIS

A search for unnatural products, that are rigid and highly-structured, and lie within the mesomolecular weight range (500-5000 Daltons) might easily lead the chemist to nominate pericyclic reactions as providing a vehicle for the construction process. Our recent synthesis[2] of a [12]cyclacene

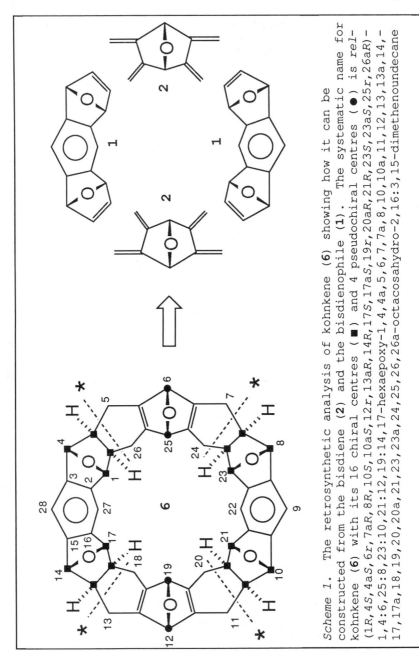

*Scheme 1.* The retrosynthetic analysis of kohnkene (**6**) showing how it can be constructed from the bisdiene (**2**) and the bisdienophile (**1**). The systematic name for kohnkene (**6**) with its 16 chiral centres (■) and 4 pseudochiral centres (●) is *rel*-(1*R*,4*S*,4a*S*,6*r*,7a*R*,8*R*,10*S*,10a*S*,12*r*,13a*R*,14*R*,17*S*,17a*S*,19*r*,20a*R*,21*R*,23*S*,23a*S*,25*r*,26a*R*)-1,4:6,25:8,23:10,21:12,19:14,17-hexaepoxy-1,4,4a,5,6,7,7a,8,10,10a,11,12,13,13a,14,-17,17a,18,19,20,20a,21,23,23a,24,25,26,26a-octacosahydro-2,16:3,15-dimethenoundecane

derivative (**6**), we have elected[3-7] to call kohnkene, supports this contention: it can be constructed[2-9] with remarkable ease (Scheme 1) as a result of four repetitive Diels-Alder reactions in two practical steps from a known[10] bisdiene (**2**) and a known[11] bisdienophile (**1**). Even although kohnkene (**6**) possesses $D_{2h}$ molecular symmetry, there are a total of 136 configurationally diastereoisomeric forms, ignoring obvious topological constraints, that could be explored by 4096 different reaction pathways. The fact that, in addition to kohnkene (**6**), two precursors with the 'correct' configurational pedigrees have been isolated, implies that *there are inherently simple ways of making apparently complex unnatural products from appropriate substrates without the need for reagent control or catalysis.*[12] Since, in principle, chemical reactions including the Diels-Alder reaction can be either kinetically or thermodynamically controlled, the *structures* of the *substrates* and/or the *products* — the latter always being a consequence of the former — can *direct* the outcome of the reaction. We could call[12] this process *structure-directed synthesis.*

During the last few years, Eschenmoser[13] has drawn attention to the concept of structure-directed synthesis in the context of natural product synthesis and the production of certain biomolecules, particularly organic cofactors. With reference to the specific structural elements, of vitamin $B_{12}$, he has commented[13] on the fact that 'these outwardly complex structural elements are found to "self-assemble" with surprising ease under structurally appropriate preconditions; the amount of "external instruction" required for their formation turns out to be surprisingly small in view of the complexity and specificity of these structural elements.' This statement follows on the back of much

experimental work[14] and, of course, the thesis[15] that the
molecular building blocks of nucleic acids, proteins, and
polysaccharides — namely the nucleic acid bases, and the
simpler α-amino acids and carbohydrates — are of prebiotic
origin. The synthesis of adenine, both thermally[16] and
photochemically,[17] from hydrogen cyanide — of which it is a
pentamer — is an earlier illustration of the structural
self-assembly of a natural product from a simple building
block.

Are there examples of such self-assembly processes
leading from simple building blocks to complex unnatural
products that compare with the formation (Scheme 1) of
kohnkene (6) from the bisdiene (2) and the bisdienophile
(1)? There are — although they lack the stereochemical
complexity of this [12]cyclacene derivative with its 16
chiral and 4 pseudochiral centres. The pumpkin-shaped
receptor, cucurbituril, is a result[18] of the acid-catalysed
self-assembly of glyoxal, urea, and formaldehyde in the
molar proportions of 6 OHCCHO:12 $H_2NCONH_2$:12 HCHO with the
loss of 24 moles of $H_2O$. The so-called calixarenes[19] and
cavitands,[20,21] built up from condensations of *para*-sub-
stituted phenols and resorcinols, free and 2-substituted,
with simple aldehydes such as formaldehyde and
acetaldehyde, are yet further examples of unnatural
products where the syntheses seem to be structure-directed.
Indeed, a recent thought-provoking analysis[22] of Cram's
carcerand[23] suggests that certain molecular topologies,
including the carcerand, are made up of atoms assembled on
a minimal surface in topological language.

In principle, there is no reason to restrict the
concept of structure-directed synthesis to organic

compounds. Structural features of self-assembly, cooperativity, and self-self recognition characterise[24,25] the new
polynuclear double-stranded helicates incorporating
poly(bipyridine) strands and copper (I) cations. Many other
metalloorganic and organometallic compounds of a highly-
structured nature will no doubt materialise in the near
future. The existence of clays and zeolites bears witness
to the spontaneous ordering of inorganic substrates.

*Constructing Molecular Belts*

    Figure 1 summarises the structural and stereo-
electronic characteristics of the bisdienophile (**1**) and
bisdiene (**2**) that have been employed[2-9] successfully in the
synthesis of kohnkene (**6**). Not only do these building
blocks have the structural characteristics of rigidity and
curvature necessary to form belt-shaped molecules, but they
are also endowed[26] with high stereochemical preferences for
various stereoelectronic reasons (see later) when they enter
into (repetitive) Diels-Alder reactions. Whilst *(i)*
dienophilic units, such as those present in (**1**), are *only*

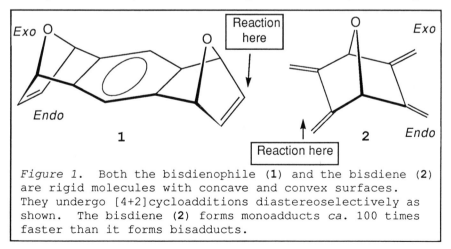

*Figure 1.*  Both the bisdienophile (**1**) and the bisdiene (**2**)
are rigid molecules with concave and convex surfaces.
They undergo [4+2]cycloadditions diastereoselectively as
shown.  The bisdiene (**2**) forms monoadducts *ca.* 100 times
faster than it forms bisadducts.

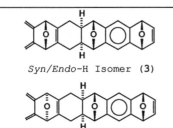

*Syn/Endo*-H Isomer (3)

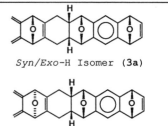

*Syn/Exo*-H Isomer (3a)

*Anti/Endo*-H Isomer (3b)

*Anti/Exo*-H Isomer (3c)

Figure 2. The four diastereoisomeric 1:1 adducts. The 'remote' stereochemical descriptors, *syn* and *anti*, refer to the relative configurations of the bridging oxygen atoms in the bicyclic rings across the newly-formed cyclohexene rings. The 'close' stereochemical descriptors, *endo* and *exo*, refer to the relative configurations of the hydrogen atoms at the ring junctions associated with the newly-created chiral centres.

attacked by dienes at their *exo* faces, *(ii)* diene units, such as those present in (2), are *only* attacked by dienophiles at their *endo* faces. The implications of *(i)* for 1:1 adduct formation in a single Diels-Alder reaction between (1) and (2) are clear: the ring junction hydrogen atoms at the newly-created chiral centres *must* adopt the *endo*-H configuration as in diastereoisomers (3) and (3b) shown in Figure 2, *i.e.* the diastereoisomeric 1:1 adducts (3a) and (3c) with their ring junction hydrogen atoms in the *exo*-H configuration will *not* be formed. In addition to these 'close' stereochemical preferences summarised in *(i)* and *(ii)* above, there is a third 'remote' stereochemical consideration when (1) undergoes a single Diels-Alder reaction with (2) to give a 1:1 adduct. That is — will the oxygen atoms bridging the 6-membered rings in the two building blocks lie on the same side (*syn*) or on different sides (*anti*) of the developing structure? The answer is that *only* the 1:1 adduct (3) with *syn* stereochemistry — and, of course, *endo*-H ring junctions — is formed for steric

reasons, *i.e.* the 'close' dictates the 'remote' stereochemistry in so far as the hypothetical transition state
leading to the *anti* 1:1 adduct **(3b)** is highly disfavoured
sterically. The treble diastereoselectivity which operates
overall during a single Diels-Alder reaction between **(1)**
and **(2)** to give the 1:1 adduct **(3)** with *syn-endo*-H
stereochemistry (Figure 2) is, to all intents and purposes,
complete. This compound, which can be isolated as a minor
product from a reaction (Scheme 2) carried out in toluene
by heating **(1)** with an excess (2 mol. equiv.) of **(2)** under
reflux (111°C) for 12 hours has been characterised (Figure
3) by X-ray crystallography of its anthracene adduct. In
addition to establishing the *syn-endo*-H configuration **(3)**
of the 1:1 adduct, the Diels-Alder reaction of the

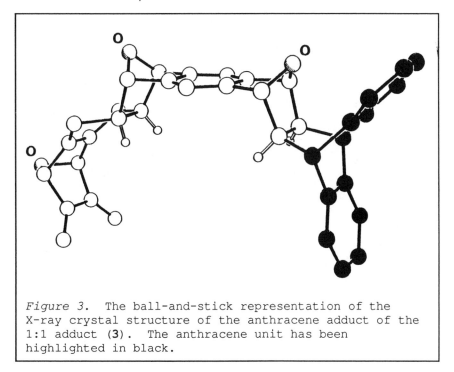

*Figure 3.* The ball-and-stick representation of the
X-ray crystal structure of the anthracene adduct of the
1:1 adduct **(3)**. The anthracene unit has been
highlighted in black.

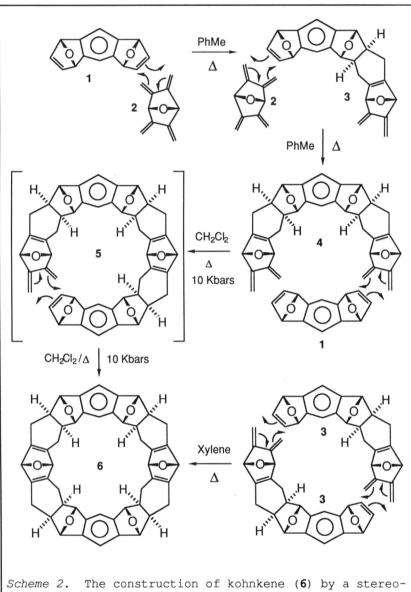

*Scheme 2.* The construction of kohnkene (**6**) by a stereo-
electronically-programmed set of Diels-Alder reactions
starting from the bisdienophile (**1**) and the bisdiene (**2**).

anthracene (the diene) at the dienophilic double bond in
(**3**) is also shown to proceed by *exo* attack to give ring
junction hydrogen atoms with *endo*-H stereochemistry as
expected. The major product (Scheme 2) which accompanies
(**3**) is the 2:1 adduct (**4**) with $C_{2v}$ molecular symmetry. It
has been *isolated* in 78% yield and its structure was
confirmed by ($^1$H and $^{13}$C) n.m.r. spectroscopy by analogy
with the spectroscopic data obtained for the 1:1 adduct
(**3**). The 2:1 adduct (**4**) accumulates in the reaction
mixture because of the much reduced reactivities of the
diene units in (**4**) towards further cycloadditions. The
difference in exothermicities associated with the consecutive Diels-Alder reactions involving the two diene units
in (**2**) is such that the second cycloaddition proceeds[10,27]
*ca.* 100 times slower than the first, on account of the
olefinic double bond introduced into the bicyclic framework. Although the dissimilar reactivities of the two
diene units in (**2**) halts the progression towards kohnkene
(**6**), it does allow a controlled and stepwise approach
towards the final product. Thus, cyclodimerisation of the
1:1 adduct (**3**) gave (**6**) in very low yield (3.5%) after 2
days heating under reflux (~140°C) in xylene. Much more
efficient was the application[28] of high pressure (10 kbars)
for 200 hours to a reaction mixture containing equimolar
amounts of the bisdienophile (**1**) and the 2:1 adduct (**4**) in
dichloromethane at 55-60°C. The best *isolated* yield
obtained so far from this particular regime is 36%.
Presumably, kohnkene (**6**) is formed *via* the intermediacy of
the acyclic 2:2 adduct (**5**), which no doubt undergoes a very
fast Diels-Alder reaction under the highly favourable
stereoelectronic conditions pertaining in the reaction.
The efficient production of kohnkene (**6**) must be a
reflection of the stereoelectronic information programmed

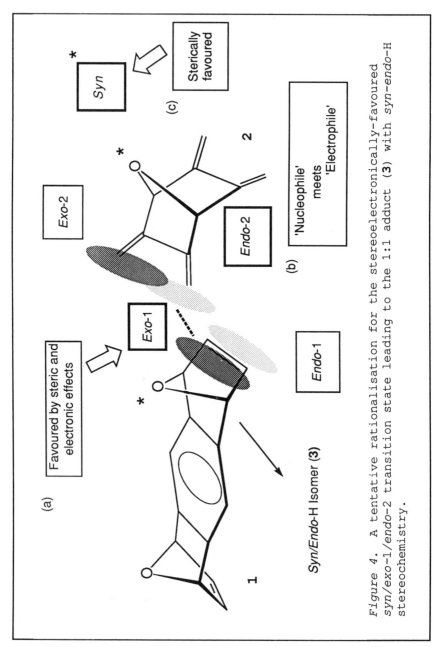

*Figure 4.* A tentative rationalisation for the stereoelectronically-favoured *syn/exo*-1/*endo*-2 transition state leading to the 1:1 adduct (**3**) with *syn–endo*-H stereochemistry.

into (1) and (2) and subsequently transmitted through (3), (4), and (5) to (6). The repetitive Diels-Alder reactions involve the following choices of diastereoisomers and reaction pathways: for (3), 1 out of 4 attainable by 8 pathways, for (4), 1 out of 10 attainable by 64 pathways, and for (5), 1 out of 64 attainable by 512 pathways, and it is this last '1', and *only* this '1', which cyclises to give (6).

Close examination (Figure 4) of the dienophilic (1) and diene (2) portions of (1) and (2) respectively has led[8] to a tentative rationalisation of the treble diastereo-selectivity that accompanies repeatedly each Diels-Alder reaction *en route* to kohnkene (6). The higher electron densities associated with the π-systems in both diene (2) and dienophilic (1) units are concentrated[26] on the *exo* faces. Intuitively, one would expect[8] reaction to occur between a face of high electron density and one of low electron density, *viz.* either an *exo*-1/*endo*-2 or an *endo*-1/

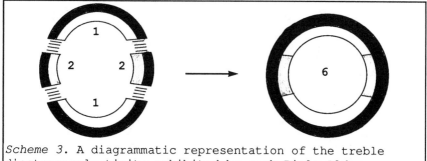

*Scheme 3.* A diagrammatic representation of the treble diastereoselectivity exhibited by each Diels-Alder reaction between two (shaded on the concave surface) bis-diene (2) and two (unshaded on the concave surface) bis-dieneophile (1) units in the construction of kohnkene (6). The convex surfaces (which are 'studded' with oxygen atoms) of the units and the product are rendered in black.

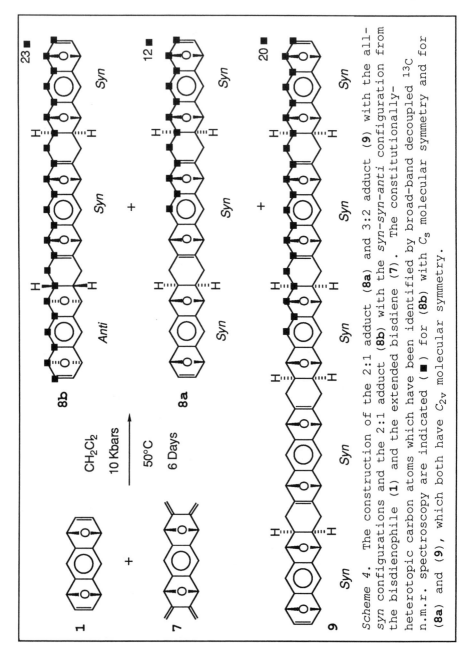

*Scheme 4.* The construction of the 2:1 adduct (**8a**) and 3:2 adduct (**9**) with the all-*syn* configurations and the 2:1 adduct (**8b**) with the *syn-syn-anti* configuration from the bisdienophile (**1**) and the extended bisdiene (**7**). The constitutionally-heterotopic carbon atoms which have been identified by broad-band decoupled $^{13}C$ n.m.r. spectroscopy are indicated (■) for (**8b**) with $C_s$ molecular symmetry and for (**8a**) and (**9**), which both have $C_{2v}$ molecular symmetry.

*exo*-2 interaction occurs at the transition state. Since
*exo*-1 attack at the dienophilic units is believed to be
favoured (Figure 4a) by relief of torsional strain in the
transition state between olefinic and bridgehead hydrogen
atoms,[26] the *exo*-1/*endo*-2 interaction occurs (Figure 4b)
leading, for example, to the observed *syn-endo*-H stereochemistry of the 1:1 adduct **(3)** when **(1)** undergoes cycloaddition with **(2)**. The 'remote' *syn* stereochemistry
(Figure 4c) which accompanies this 'close' double
diastereoselectivity, is a direct steric consequence. This
kind of transition state molecular recognition is repeated
(Scheme 3) four times over in the construction of kohnkene
**(6)** from its 'intelligent' substrates **(1)** and **(2)**. The
shapes and rigidity of all the components and the nature of
their mutual fits dictate the course and net result of this
structure-directed synthesis of a molecular belt. Kohnkene
**(6)** has a ring of molecular 'LEGO' about it.

*Constructing Molecular Coils*

How general is the concept of structure-directed
synthesis when applied to Diels-Alder reactions between
different bisdienes and bisdienophiles? Can the oxygen
atom bridges in **(1)** and **(2)** be replaced by other
heteroatoms and groups with retention of the treble
diastereoselectivity? When different 'spacer' groups are
introduced into the bisdienes and bisdienophiles, altering
the rigidity and/or the curvature of the building blocks,
what are the consequences for molecular belt formation?

Under high pressure, the (*syn*)-bisdienophile **(1)**
reacts (Scheme 4) with the extended[29] *syn*-bisdiene **(7)** in
dichloromethane solution to afford the 2:1 adduct **(8a)** and

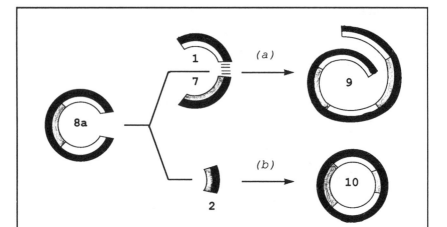

*Scheme 5.* A diagrammatic representation of the
construction by repetitive Diels-Alder reactions of a
molecular coil in the shape of the 3:2 adduct (**9**) or a
molecular belt in the form of the [14]cyclacene
derivative (**10**), assuming the presence of the 2:1
adduct (**8a**) in admixture *(a)* with the extended
bisdiene (**7**) and the bisdienophile (**1**) or *(b)* with the
'smaller' bisdiene (**2**). The concave surfaces of
bisdiene units are shaded whereas those of
bisdienophilic units are not. The convex surfaces
(which are 'studded' with oxygen atoms) of the units
separately and in the starting material and the
products are rendered in black.

the 3:2 adduct (**9**) as the major (25%) and minor (2%)
products respectively. The all-*syn* configurations of
these adducts, containing 11 and 19 laterally-fused
6-membered rings respectively, has been established[12]
unambiguously by ($^1$H and $^{13}$C) n.m.r. spectroscopy. Once
again, they are a result (Scheme 5) of trebly diastereo-
selective Diels-Alder reactions, involving exclusively
*endo* and *exo* approaches, respectively, with reference to
the diene and dienophilic units. The 2:1 and 3:2 adducts

contain 16 and 28 chiral centres, respectively, and can exist in 10 and 136 configurationally-diastereoisomeric forms. However, on this occasion, the all-*syn* 2:1 adduct (**8a**) is accompanied by trace amounts (1 part in 16) of the *syn-syn-anti* 2:1 adduct (**8b**): this implies the existence of a minor reaction pathway where the approach is *exo* with respect to both diene and dienophilic units in one of the two cycloadditions between (**1**) and (**7**). Interestingly, it has not been possible to isolate a [16]cyclacene derivative despite numerous attempts to react the all-*syn* 2:1 adduct (**8a**) with the extended *syn*-bisdiene (**7**). We believe that the stereoelectronic requirements for cyclisation are not being fulfilled and that the growing oligomer prefers to form (Scheme 5) a molecular coil, wherein the potentially reactive end groups cannot react intramolecularly. This hypothesis is supported by appreciable shifts to lower frequencies of the signals in the $^1$H n.m.r. spectrum for protons sited near the ends of the oligomeric chain comprising the 3:2 adduct (**9**). This observation suggests, along with the inspection of molecular models, that the 3:2 adduct (**9**) with its all-*syn* configuration adopts (Scheme 5) a 'Swiss-roll-like' conformation. The experiment, anticipated diagramatically in Scheme 5, of employing a 'smaller' bisdiene to effect cyclisation with (**8a**) is summarised in Scheme 6. Quite remarkably, the 2:1 adduct (**8a**) undergoes cyclisation with the bisdiene (**2**) to give the [14]cyclacene derivative (**10**) in 76% yield *without having to resort to the use of high pressure*. The fact that this reaction proceeds in toluene heated under reflux indicates that the final intramolecular Diels-Alder cyclisation must be highly favoured stereoelectronically.

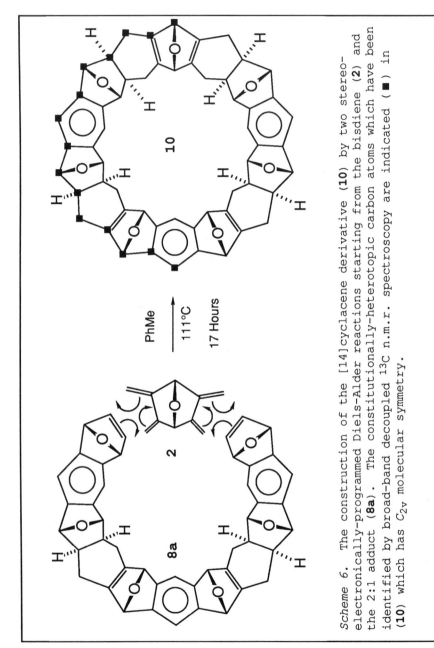

*Scheme 6.* The construction of the [14]cyclacene derivative (**10**) by two stereo-electronically-programmed Diels-Alder reactions starting from the bisdiene (**2**) and the 2:1 adduct (**8a**). The constitutionally-heterotopic carbon atoms which have been identified by broad-band decoupled $^{13}C$ n.m.r. spectroscopy are indicated (■) in (**10**) which has $C_{2v}$ molecular symmetry.

Constructing Molecular Cages

The bisdienophile (1), employed with appropriate
bisdienes to construct belts (Schemes 1, 2, and 6) and
coils (Scheme 4) contains two dienophilic units fused
diagonally on to a benzene ring. One can equally well
envisage[30] using (Scheme 7) the known[31] trisdienophile with
the all-*syn* configuration (11), on account of the trigonal
disposition of the three dienophilic units around a
central benzene ring, to construct molecular cages with
bisdienes such as (2). As in the case of kohnkene (6),
the synthesis can be performed in a stepwise manner by,
first of all, relying on a thermally-promoted
cycloaddition to add three bisdienes (2) to the
trisdienophile (11) to give the trisadduct (12) and then,
by resorting to high pressure reaction conditions, to cap
the trisadduct (12) with a second trisdienophile (11).
Using 3.3 molar equivalents of the bisdiene (2), the first
reaction proceeds under reflux in toluene to give the
trisadduct (12), which was *isolated* in 48% yield by
chromatography from minor amounts of a bisadduct and a
monoadduct. Once again, we witness a remarkable
sterochemical event. Although 16 configurationally
diastereoisomeric trisadducts could, in principle, be
formed, only *one* has been isolated so far. [1]H N.m.r.
spectroscopy established not only the $C_{3v}$ symmetry of the
trisadduct (12) but also the *endo* configurations for the
methine hydrogen atoms at the junctions of the three
newly-formed cyclohexene rings. The cage compound (13),
which we have called[30] trinacrene after an old name
(Trinacria) for Sicily, was obtained (Scheme 7) by
subjecting equimolar amounts of the trisadduct (12) and
the trisdienophile (11) to high pressure in

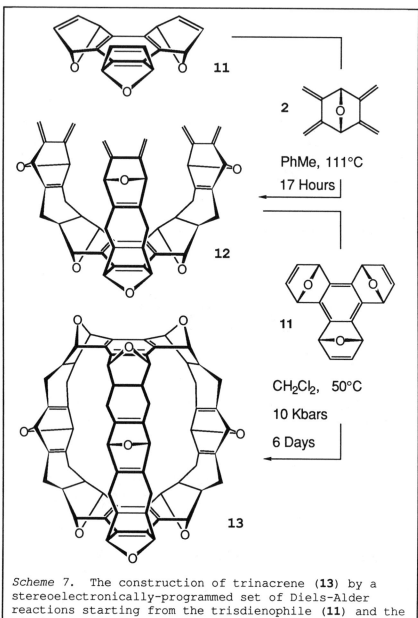

*Scheme* 7. The construction of trinacrene (**13**) by a stereoelectronically-programmed set of Diels-Alder reactions starting from the trisdienophile (**11**) and the bisdiene (**2**).

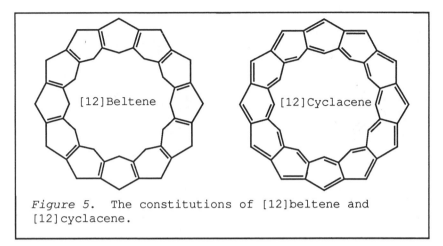

*Figure 5.* The constitutions of [12]beltene and [12]cyclacene.

dichloromethane solution. Presumably, an intermolecular Diels-Alder reaction is followed by two intramolecular cycloadditions to afford trinacrene (**13**) with its 24 chiral and 6 pseudochiral centres. This cage construction represents yet another convincing demonstration of structure-directed synthesis at work.

*Conclusions*

Kohnkene (**6**) and the related [14]cyclacene derivative (**10**) represent the first two key compounds for the elaboration of a whole series of exotic hydrocarbons, including the [n]beltenes,[32] or [n]columnenes,[33] the [n]collarenes,[3] and the [n]cyclacenes.[34-36] Using the clock numbering system[3] employed in Scheme 8, note how deoxygenation[37] can be effected[3] at 3 and 9 o'clock to give dideoxykohnkene (**14**), followed by dehydration[11] at 1, 5, 7, 11 o'clock to afford[3] the unsymmetrical (*i.e.* disordered) hydrocarbon (**15**), containing one anthracene, one benzene, and two naphthalene rings. The well-known[38]

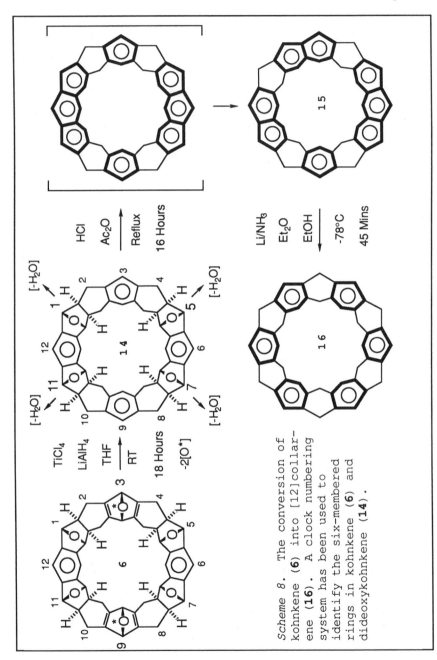

*Scheme 8.* The conversion of kohnkene (6) into [12]collarene (16). A clock numbering system has been used to identify the six-membered rings in kohnkene (6) and dideoxykohnkene (14).

acid-catalysed migration of aromatic rings in acyclic compounds is also, not surprisingly, observed in polymacrocyclic arrays. Are entropic considerations dictating the outcome of this reaction? It seems likely they are. Whilst the hydrocarbon (15) could be dehydrogenated to give [12]cyclacene (Figure 5), it could equally well be hydrogenated to afford [12]beltene (Figure 5). So far, an attempted Birch reduction of (15) has led[3] to the isolation of [12]collarene (16) with its six alternating benzene rings in a cyclic array. Similar structural modifications could be made to the 2:1 adduct (8a) and the 3:2 adduct (9) to provide building blocks for 'ladder' polymers.[38-40] Trinacrene (13) is amenable to structural modifications similar to those described in Scheme 8 for kohnkene (6).

Compounds resembling molecular belts, coils, and cages could also possess intriguing receptor properties: dideoxykohnkene (14) crystallises[3] with a lone water molecule in its middle and [9]beltene has been predicted,[32] from molecular mechanics calculations, to complex with acetylene: computer molecular graphics indicate[30] that the cavity of trinacrene (13) could incorporate a wide variety of guest species, both neutral and charged.

It is the Diels-Alder reaction[41] — employed repet-itively — in structure-directed synthesis that has afforded the rigid, ordered, and large precursors to these new and novel hydrocarbons. One cannot help but reflect on the spontaneity and evolutionary characteristics of the Diels-Alder chemistry involving rigid and ordered bisdienes (*e.g.* 2 and 7), bisdienophiles (*e.g.* 1), and trisdienophiles (*e.g.* 11). They serve to illustrate the thesis advanced by

Atkins[42] a few years ago which went.. .. 'Small molecules
evolve into bigger ones by eating smaller ones, although it
is not always clear which is the eater and which the eaten.
Little molecules eat by impact, and what emerges from their
collisions is sometimes a bigger molecule' ...'Complex
molecules emerge from simpler ones inhabiting planets; and
so the specification could be simplified still further.
That simplification can itself be simplified, for if you
merely specify the elements, and perhaps some other things,
sooner or later there will be elephants.'

There is another fascinating feature about the Diels-
Alder reaction, one of the most powerful of the key ring-
forming reactions[43] in synthetic organic chemistry.  As
Breslow[44] has pointed out, 'The Diels-Alder reaction is one
of the most useful synthetic processes, although it does
not play an important role in biochemistry.  In fact, the
evidence that Diels-Alder reactions occur in nature at all
simply comes from the consideration of structures of some
natural products which look as if they must have been
derived from such reactions, and no characterized Diels-
Alder-catalysing enzymes have yet been described.'  The
very minor role of the Diels-Alder reaction, as one of the
few common organic reactions without numerous enzymic
equivalents, in biosynthetic pathways is puzzling.  Nature
is usually reluctant to miss a chance.  However, in this
case, it might just be that there is enough 'chemical'
selectivity associated with those Diels-Alder reactions
that occur in nature for it to obviate the need for enzymic
control.  Where natural products (*e.g.* racemic endiandric
acid[45] and the alkaloids,[46] catharanthine and tabersonine)
are probably formed by Diels-Alder reactions, it is not
unlikely that a *nonenzymatic cascade mechanism* [47] comes

into play. It seems, however, that selectively-raised abzymes[49] might offer an opportunity[50] to catalyse the Diels-Alder reaction, since 'antibodies raised against structures which mimic the transition states of chemical reactions should selectively catalyse those reactions'.

It is tempting to speculate that the science of the structure-directed synthesis of unnatural products might best be developed around those chemical reactions for which nature has not found it necessary to evolve and involve enzymes.

**ACKNOWLEDGEMENTS**

We are most grateful to Dr. Neil S. Isaacs (University of Reading) for performing all the high pressure reactions, to Mr. Peter R. Ashton and Dr. Neil Spencer (University of Sheffield) for recording fast atom bombardment mass spectra and nuclear magnetic resonance spectra, respectively, and to Dr. David J. Williams and Miss Alexandra M.Z. Slawin (Imperial College London) for carrying out the X-ray crystal structure determinations. We thank Prof. Dr. R.W. Saalfrank (Erlanger-Nurnberg) for discussions on the use of the term 'Strukturegerechte Synthese', Dr. R. Neier (Université de Fribourg Suisse) and Dr. I. Markó (University of Sheffield) for drawing our attention to examples of the occurrence of Diels-Alder reactions in the formation of natural products, and Dr. R.L. Baxter (University of Edinburgh) for identifying a reference to the Diels-Alder reaction in the recent literature on abzymes. The research was supported by the University of Sheffield and the Science and Engineering Research Council and the Ministry of Defence in the United Kingdom, and the University of Messina in Italy. One

266 *Molecular Recognition*

(J.F.S.) of us acknowledges the award of a Research Fellowship from the Leverhulme Trust.

**REFERENCES**

1. S.L. Miller and L.E. Orgel, 'The Origin of Life on
   the Earth,' Prentice-Hall, Englewood Cliffs, New
   Jersey, USA, 1974.
2. F.H. Kohnke, A.M.Z. Slawin, J.F. Stoddart, and D.J.
   Williams, *Angew. Chem. Int. Ed. Engl.*, 1987, *26*, 892.
3. P.R. Ashton, N.S. Isaacs, F.H. Kohnke, A.M.Z. Slawin,
   C.M. Spencer, J.F. Stoddart, and D.J. Williams,
   *Angew.Chem. Int. Ed. Engl.*, 1988, *27*, 966.
4. J.F. Stoddart, *Nature*, 1988, *334*, 10.
5. J.F. Stoddart, *Chem. Brit.*, 1988, *24*, 1203.
6. J.F. Stoddart, *J. Incl. Phenom.*, 1989, 7, 000 (In
   press).
7. L. Milgrom, *New Scientist*, No. 1641, 3 Dec. 1988,
   p.61.
8. P. Ellwood, J.P. Mathias, J.F. Stoddart, and F.H.
   Kohnke, *Bull. Soc. Chim. Belg.*, 1988, *97*, 669.
9. F.H. Kohnke and J.F. Stoddart, *Pure Appl. Chem.*,
   1989, *61*, 000 (In press).
10. P. Vogel and A. Florey, *Helv. Chim. Acta*, 1974, *57*,
    200; C. Mahaim, P.-A. Carrupt, J.-P. Hagenbuch, A.
    Florey, and P. Vogel, *Helv. Chim. Acta*, 1980, *63*,
    1149; F.H. Kohnke, J.F. Stoddart, A.M.Z. Slawin, and
    D.J. Williams, *Acta. Cryst.*, 1988, *C44*, 736.
11. H. Hart, N. Raja, M.A. Meador, and D.J. Ward, *J. Org.
    Chem.*, 1983, *48*, 4357; F.H. Kohnke, J.F. Stoddart,
    A.M.Z. Slawin, and D.J. Williams, *Acta Cryst.*, 1988,
    *C44*, 738, 742.
12. P.R. Ashton, N.S. Isaacs, F.H. Kohnke, J.P. Mathias,
    and J.F. Stoddart, *Angew. Chem. Int. Ed. Engl.*,
    Submitted.
13. A. Eschenmoser, *Angew. Chem. Int. Ed. Engl.*, 1988,
    *100*, 5.
14. G. Kansder, G. Bold, R. Lattman, C. Lehmann, T. Früh,
    Y.-B. Xiang, K. Inomata, H.-P. Buser, J. Schreiber,
    E. Zass, and A. Eschenmoser, *Helv. Chim. Acta*, 1987,
    *70*, 1115.

15. S.L. Miller, *Science*, 1953, *117*, 528; *J. Am. Chem. Soc.*, 1955, 77, 2351; *Biochem. Biophys. Acta*, 1957, *23*, 480.

16. J. Oró, *Biochem. Biophys. Res. Commun.*, 1960, *2*, 407; J. Oró and B.P. Kimball, *Arch. Biochem. Biophys.*, 1961, *64*, 217; 1962, *96*, 293.

17. J.P. Ferris and L.E. Orgel, *J. Am. Chem. Soc.*, 1966, *88*, 1074.

18. W.A. Freeman, W.L. Mock, and N.-Y. Shih, *J. Am. Chem. Soc.*, 1981, *103*, 7367; W.L. Mock and N.-Y. Shih, *J. Org. Chem.*, 1983, *48*, 3618; 1986, *51*, 4440; W.L. Mock, T.A. Mirra, J.P. Wipsiec, and T.L. Manimaran, *J. Org. Chem.*, 1983, *48*, 3619; W.L. Mock and N.-Y. Shih, *J. Am. Chem. Soc.*, 1988, *110*, 4706; 1989, *111*, 2697.

19. C.D. Gutsche, *Acc. Chem. Res.*, 1983, *16*, 161; *Top. Curr. Chem.*, 1984, 123, 1; *Prog. Macrocyclic Chem.*, 1987, *3*, 93.

20. D.J. Cram, *Science*, 1983, *219*, 1177; *1988*, 240, 760; *Angew. Chem. Int. Ed. Engl.*, 1986, *25*, 1039; 1988, *27*, 1009.

21. D.J. Cram, K.D. Stewart, I. Goldberg, and K.N. Trueblood, *J. Am. Chem. Soc.*, 1985, *107*, 2574; D.J. Cram, S. Karbach, H.E. Kim, C.B. Knobler, E.F. Maverick, J.L. Ericson, and R.C. Helgeson, *J. Am. Chem. Soc.*, 1988, *110*, 2229.

22. Z. Blum and S. Lidin, *Acta Chem. Scand.*, 1988, *B42*, 332.

23. D.J. Cram, S. Karbach, Y.H. Kim, L. Baczynskyj, and G.W. Kalleymeyn, *J. Am. Chem. Soc.*, 1985, *107*, 2575; D.J. Cram, S. Karbach, Y.H. Kim, L. Baczynskyj, K. Marti, R.M. Sampson, and G.W. Kalleymeyn, *J. Am. Chem. Soc.*, 1988, *110*, 2554.

24. J.-M. Lehn, A. Rigault, J. Siegel, J. Harrowfield, B. Chevier, and D. Moras, *Proc. Nat. Acad. Sci. USA*, 1987, *84*, 2565; J.-M. Lehn and A. Rigault, *Angew. Chem. Int. Ed. Engl.*, 1988, *27*, 1095.

25. J.-M. Lehn, *Angew. Chem. Int. Ed. Engl.*, 1988, *27*, 89.

26. W.H. Watson (Ed.), 'Stereochemistry and Reactivity of Systems containing $\pi$-Electrons,' Verlag Chemie, Deerfield Beach, Florida, 1983: in particular, the articles by K.N. Houk (p.1), L.A. Paquette (p.41), R. Gleiter and M.C. Böhm (p.105), and P. Vogel, (p.147).

27. Y. Bessière and P. Vogel, *Helv. Chim. Acta*, 1980, *63*, 232.

28. N.S. Isaacs and A.V. George, *Chem. Brit.*, 1987, *23*, 47.

29. J. Luo and H. Hart, *J. Org. Chem.*, 1988, *53*,1341; 1989, *54*, 1762; F.H. Kohnke, J.P. Mathias, J.F. Stoddart, A.M.Z. Slawin, D.J. Watts, and D.J. Williams, *Acta Cryst.*, Submitted.

30. P.R. Ashton, N.S. Isaacs, F.H. Kohnke, G. Stagno d'Alcontres, and J.F. Stoddart, *Angew. Chem. Int. Ed. Engl.*, Submitted.

31. M.B. Stringer and D. Wege, *Tetrahedron Lett.*, 1980, *21*, 3831.

32. R.W. Alder and R.B. Sessions, *J. Chem. Soc., Perkin Trans. 2*, 1985, 1849; A. Nickon and E.F. Silversmith, 'Organic Chemistry: The Name Game', Pergamon Press, New York, 1987, p.110.

33. R.O. Angus, Jr. and R.P. Johnson, *J. Org. Chem.*, 1988, *53*, 314.

34. A.T. Balaban, *Pure Appl. Chem.*, 1980, *52*, 1409; *Rev. Roumaine Chim.*, 1988, *33*, 699.

35. F. Vögtle, *Top. Curr. Chem.*, 1983, *115*, 157.

36. S. Kivelson and O.L. Chapman, *Phys. Rev.*, 1983, *B28*, 7235.

37. H. Hart and G. Nwokoga, *J. Org. Chem.*, 1981, *46*, 1251; Y.D. Xing and N.Z. Huang, *J. Org. Chem.*, 1982, *47*, 140; H.N.C. Wong, *Acc. Chem. Res.*, Submitted.

38. E. Clar, 'Polycyclic Hydrocarbons', Academic Press, New York, 1964, Vol. 1, p.64; J. Luo and H. Hart, *J. Org. Chem.*, 1987, *52*, 4833.

39. L.L. Miller, A.D. Thomas, C.L. Wilkins, and D.A. Weil, *J. Chem. Soc., Chem. Commun.*, 1986, 661; A.D. Thomas and L.L. Miller, *J. Org. Chem.*, 1986, *51*, 4160; W.C. Christopfel and L.L. Miller, *J. Org. Chem.*, 1986, *51*, 4169, W.C. Christopfel and L.L. Miller, *Tetrahedron*, 1987, *43*, 3681; T. Chiba, P.W. Kenny, and L.L. Miller, *J. Org. Chem.*, 1987, *52*, 4327; P.W. Kenny and L.L. Miller, *J. Chem. Soc., Chem. Commun.*, 1988, 84; P.W. Kenny, L.L. Miller, S.F. Rak, T.H. Jozekiak, W.C. Christopfel, J.-H. Kim, and R.A. Uphaus, *J. Am. Chem. Soc.*, 1988, *110*, 4445.

40. J. Luo and H. Hart, *J. Org. Chem.*, 1987, *52*, 3631.

41. O. Diels and K. Alder, *Ann. Chem.*, 1928, *460*, 98.

42. P.W. Atkins, 'The Creation', Freeman, Oxford, 1981, p. 3 and 4.

43. E.J. Corey, W.J. Howe, and D.A. Pensak, *J. Am. Chem. Soc.*, 1974, *96*, 7724.

44. R. Breslow, in 'Inclusion Compounds', eds. J.L. Atwood, J.E.D. Davies, and D.D. MacNicol, Academic Press, London, 1984, Vol. 3, p. 473.

45. W.M. Bandaranayake, J.E. Banfield, and D.St.C. Black, *J. Chem. Soc., Chem. Commun.*, 1980, 902.

46. D.R. Dalton in 'Studies in Organic Chemistry. The Alkaloids: The Fundamental Chemistry — A Biogenetic Approach', ed. P.G. Gassman, Dekker, 1979, Vol. 7, Ch. 30, p. 432.

47. K.C. Nicolaou and N.A. Petasis, in 'Strategies and Tactics in Organic Synthesis', ed. T. Lindberg, Academic Press, Orlando, Florida, 1984, Ch. 6, p. 155.

48. L. Pauling, *Chem. Eng. News*, 1946, *24*, 1375; *Am. Scientist*, 1948, *36*, 51; W.P. Jencks, 'Catalysis in Chemistry and Enzymology', McGraw-Hill, New York, 1969, p. 288.

49. A. Tramontano, K.D. Janda, and R.A. Lerner, *Science*, 1986, *234*, 1566; S.J. Pollack, J.W. Jacobs, and P.G. Schultz, *Science*, 1986, *234*, 1570; J. Jacobs, P.G. Schultz, R. Sugasawara, and M. Poweli, *J. Am. Chem. Soc.*, 1987, *109*, 2174; A.D. Napper, S.J. Benkovic, A. Tramontano, and R.A. Lerner, *Science*, 1987, *237*, 1041.

50. A. Balan, B.P. Doctor, B.S. Green, M. Torten, and H. Zuffer, *J. Chem. Soc., Chem. Commun.*, 1988, 106.

# The Physical-organic Chemistry of Surfaces, and its Relevance to Molecular Recognition

By G. M. Whitesides and Hans Biebuyck

DEPARTMENT OF CHEMISTRY, HARVARD UNIVERSITY, CAMBRIDGE, MASSACHUSETTS, MA 02138, USA

Molecular recognition in most biological systems is the result of summing large numbers of small, compensating energies between the parts of the interacting components. In biological systems, there are, as a rule, three major interacting components -- a protein (a receptor or enzyme), a ligand (the molecular being recognized), and water (Figure 1). Other participating molecules -- ions, membranes, small organic molecules, additional ligands -- may be important in some circumstances, but can, as a first approximation, be neglected.

How should one think about the interactions between these three major participants? Protein-water and ligand-water interactions should be easier to understand than protein-ligand interactions, since water is a homogeneous medium (or at least more homogeneous than protein or ligand). Even for interactions involving water, however, the intuition of the chemist or biochemist is less than perfect. In principle, and in due course, all will be calculated by the computer, and advances in molecular mechanics and molecular dynamics have been rapid in the last years.[1] The fact remains, however, that at the present time most of the success in practical applications of molecular recognition -- especially to drug design and molecular pharmacology -- have been based on the time-honored patterns of classical medicinal chemistry.

We have been interested in a problem whose relevance to molecular recognition apparently lies at the farthest boundaries of the subject -- that is, the physical-organic chemistry of wetting. Our interest is to understand how fluids, especially water, interact with

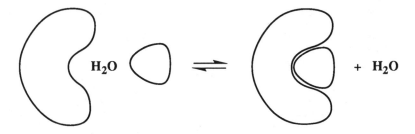

**van der Waals**
**H-Bond**
**Electrostatic**
**Water**
**Conformational Change**
$\Delta G = \Delta H - T\Delta S$

Figure 1.  A schematic illustration of the interaction
between a protein and a ligand in water.

solids or immiscible liquids.[2]  In brief, what molecular
interactions determine the energies of interfaces --
liquid-solid, solid-vapor, and liquid-vapor?  These
configurations are relevant to molecular recognition and
receptor-ligand binding, because the same interactions
determine wetting and recognition.  In wetting, however,
the systems are structurally very simple -- in principle
no more than one molecularly homogeneous, smooth liquid
interacting with an equally homogeneous, smooth surface
(of a solid or liquid).  In molecular recognition, by
contrast, the interacting components are rough (on a
molecular scale) and heterogeneous.  Nonetheless, by
understanding the interaction of a large, homogeneous,
model surface with water, we should be able to help to
understand the interaction of a small patch of protein
having similar physical characteristics with water.  By
examining a number of surfaces modelling different parts
of the protein, we may be able to contribute to
understanding the global interactions of a protein (or a
ligand) with water, and, ultimately, of a protein with a
ligand.  Conversely, if we <u>cannot</u> understand the
interaction of homogeneous surfaces with pure water, our
ability to understand the much more complex interactions
occurring in biological systems will remain unsatisfactory
at any level deeper than empirical modelling.

     In order to study solid-liquid interactions using the
approaches of physical-organic chemistry, we must be able
to prepare surfaces having well-defined structures, to
vary the structures of these surfaces, and to measure one
or more properties of the system that give information
about the energetics of the interfaces of interest.  Our
work has been addressed primarily to two subjects --
development of methods for the preparation of structurally
well-defined organic surfaces using a group of techniques
called collectively "molecular self-assembly," and the
inference of liquid-solid interactions through the
measurement of contact angles.[3,4]

     <u>Self-Assembled Monolayers</u>.  The principles of self-
assembly are well illustrated by the best developed and
most thoroughly investigated of the systems that form
organic monolayers -- long-chain alkylthiols $HS(CH_2)_nX$
adsorbed on gold.  In this technique, one exposes a clean
gold film to a solution or vapor of an organic thiol
(Figure 2).[5]  The sulfur atoms of the thiol coordinate to
the gold and are converted (by presently undefined
mechanisms) to gold thiolates having a geometry determined
by the local coordination chemistry of the sulfur and
gold.  The alkyl chains arrange themselves to minimize

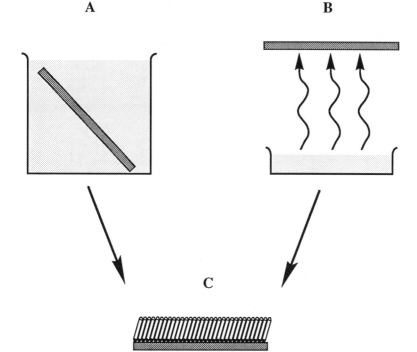

Figure 2.   There are two approaches to forming self-
            assembled monolayers on gold.  In scheme **A**, a
            gold film is immersed in a dilute solution of
            an alkanethiol; film formation is typically
            complete, as judged by macroscopic properties
            of the film, in one or two minutes.  In scheme
            **B**, the gold film is exposed to an alkanethiol
            in the gas phase, either in air or under vacuum
            conditions.  Both methods of preparation result
            in crystalline films with a cant angle of $30°$,
            **C**.  Film formation is driven by the strong,
            specific interaction of the thiol with the gold
            substrate.

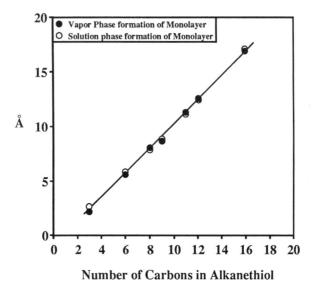

**Number of Carbons in Alkanethiol**

Figure 3.   The thickness of the film on gold is easily
controlled by changing the length of the
alkanethiol.   The average thickness of the film
in angstroms, derived from the measurement of
the intensity of the photoelectron signal of
gold using an X-ray photoelectron
spectrophotometer (XPS), is plotted as a
function of total number of carbons in the
alkanethiol.   Films were made either by
exposing the substrate to a vapor of the
alkanethiol in air (black dots) or to a
solution of the alkanethiol in ethanol (white
dots).   The line represents the best fit of the
two sets of data; the slope of this line shows
that the thickness of the film changes by
approximately an angstrom per carbon in the
alkanethiol.   Techniques used to infer the
thickness of a monolayer from the photoelectron
intensity measured by XPS are given in C.D.
Bain and G.M. Whitesides, J. Phys. Chem, 1989,
93, 1670.

their energy. The resulting systems can be shown to be
monolayers having (at least locally) well-defined
crystalline order.[6]

The advantages of this preparative technique
(relative to Langmuir-Blodgett procedures or to the
functionalization of pre-existing solids) from the vantage
of organic surface chemistry are many. The monolayers are
under thermodynamic control, and the order that they
exhibit represents a minimum in energy. They are
relatively free of defects. They will form on any
appropriately exposed surface, even on the insides of
objects and on very rough or porous surfaces. Assembly is
easy -- the gold surface is simply exposed to the thiol
for one or two hours at room temperature, removed, and
washed. Most importantly, molecular self-assembly
provides great control over certain of the physical
properties of the monolayers. In particular, the
thickness of the monolayer can be controlled in
approximately one-angstrom steps by varying the length of
the polymethylene chains (Figure 3), and the structure of
the solid-liquid interface can be controlled by varying
the terminal group X.

Only some of the rules for building well-ordered
monolayers are presently known. In particular, the
question of the influence of incommensurate sizes of the
thiolate head groups and the tail group X on the structure
of the monolayer is not well understood, and almost all
studies have used thiols of the structure $HS(CH_2)_nX$.
Nonetheless, for relatively small organic functional
groups (X = F, Cl, Br, I, CN, $CO_2H$, $CONH_2$, etc.) these
monolayers constitute readily available systems that
present sheets of organic functionality to liquid water.
By working with an extended range of thiols $HSRX(Y,Z...)$,
both as pure compounds and as mixtures, we believe that it
will eventually be possible to make a very broad range of
surfaces.

_Wetting and Contact Angle_. We have explored the
interaction between solid and liquid by measuring contact
angles (Figure 4). This technique has three great
advantages: First, it is experimentally very
straightforward (at least for static drops). Second, it
is very sensitive to the structure and composition of the
outermost part of the surface. We will show below that
the water "senses" the solid to a depth less than 10 Å,
and probably less than 5 Å -- a depth sensitivity greater
than that of many of the much more complex techniques of
vacuum physics. Third, it is applicable to the

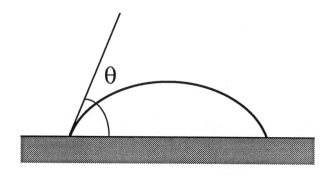

$$\gamma_{LV}\cos(\theta) = \gamma_{SV} - \gamma_{SL}$$

Figure 4.  The figure shows the equilibrium shape of a
           drop of a liquid on a flat, solid surface.  The
           contact angle of the liquid is the angle
           between the surface and the line drawn tangent
           to the sphere of liquid where the liquid meets
           the surface.  Young's equation describes the
           relationship between the contact angle of the
           liquid and the surface tensions between each of
           the three phases (solid, liquid, and gas),
           where $\gamma_{SL}$, $\gamma_{SV}$, and $\gamma_{LV}$ are the solid-liquid,
           solid-vapor, and liquid-vapor surface tensions.

exploration of solid-_liquid_ interfaces and is not
restricted to solid-vapor (vacuum) interfaces. It suffers
from two corresponding disadvantages. It is a macroscopic
rather than a molecular technique -- that is, it measures
a property determined by the collective behavior of a
large number of molecules ($10^{12}$-$10^{13}$ organic functional
groups occupy the interface between solid and liquid for a
drop of volume 1 $\mu$L). It also measures a _ratio_ of surface
energies rather than any single energy. In practice, the
liquid-vapor interfacial free

$$\cos\,\theta = (\gamma_{SV} - \gamma_{SL})/\gamma_{LV}$$

energy (that is, the surface tension) is well known, so
that the property measured directly is the difference
between the solid-vapor and solid-liquid free energies.
Varying the properties of the contacting liquid -- for
example the size of its constituent molecules or its
hydrophilicity -- can provide some insight into the nature
of the forces due to the solid.

    <u>Single-Component Monolayers -- The Depth Sensitivity</u>
<u>of Wetting</u>.  We have used monolayers comprising a single
component to investigate a number of phenomena. One of
relevance to molecular recognition is the depth
sensitivity of wetting. How far into a solid does a
liquid "see"?

    Two experiments illustrate the approach that we have
used in studying this question.  In one, we prepared
monolayers of a series of thiols $HS(CH_2)_{11}O(CH_2)_nCH_3$ in
which a polar oxygen functionality was positioned at a
controlled (by n) distance from the solid-liquid
interface, and examined the wettability of these
monolayers as a function of the carbon number n. We found
that[7] the influence of the length of the chain became
constant for n > ~6 (Figure 5).  That is, for n > 6, the
wettability of the ether-containing monolayer surface was
the same as the wettability of a pure hydrocarbon
monolayer (such as that from $HS(CH_2)_{17}CH_3$).  We conclude
that the liquid water was able to sense the "buried" ether
oxygen (by whatever mechanism -- dipole-dipole interaction
through the superficial alkyl groups or penetration of
water into the monolayer) only through less than 5-10 Å of
alkane.

    In the second experiment, we examined the wettability
of monolayers of simple alkyl thiols $HS(CH_2)_nCH_3$ on gold.
We found[5] that the wettability of these monolayers also
began to increase for n < 10 (Figure 6).  Again, the

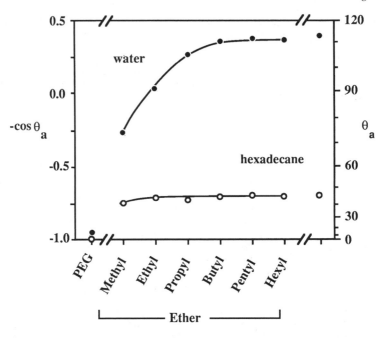

Figure 5.   Films on gold formed from ω-mercaptoethers, $HS(CH_2)_{16}O(CH_2)_nCH_3$, have interfacial properties that change as the polar ether group is buried underneath the interface.  The advancing contact angles of water and hexadecane are plotted as a function of n, the number of methylenes between the ether and the interface.  Also shown are the contact angles on polyethylene glycol (PEG) and on a monolayer of docosanethiol on gold.  The former surface is one in which the ether groups are fully exposed to the contacting liquid; the latter is one in which there is no contribution of an ether to the interfacial properties of the surface.

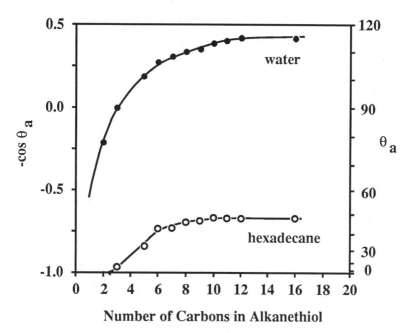

Figure 6. The contact angles on monolayers on gold formed from alkanethiols show progressively less dependence on the underlying gold substrate as the length of the alkanethiol increases. The advancing contact angles of water and hexadecane are plotted as a function of the number of carbons in the alkanethiol. The line through the data is included as a guide for the eye.

experiment does not distinguish between dipole-induced
dipole interactions involving the water and the highly
polarizable gold substrate from interactions resulting
from penetration of water between the alkyl chains down to
the gold.  In either event, these experiments confirm the
conclusion that wetting is a short-range interaction.  In
terms relevant to molecular recognition, the interaction
of water with a protein or ligand is dominated by the
structure of the outermost few angstroms of these
molecules.

Related experiments involving functional groups
positioned at the surface of polyethylene provide
independent confirmation and illustration of the short-
range character of wetting.[8]

Mixed Monolayers.  Our most useful experimental
systems are those based on monolayers incorporating two
different components, $HS(CH_2)_mX$ and $HS(CH_2)_nY$.  By varying
X and Y for chains of the same length ($m = n$), one can
investigate interactions between X and Y.[9]  By making the
chains of different lengths ($m \neq n$), one can introduce
controlled amounts of disorder into the outermost parts of
the monolayer.[10,11]

One type of experiment based on mixed monolayers is
particularly relevant to molecular recognition.  We
prepared monolayers containing a mixture of n-alkyl thiols
having two different lengths, and examined the wetting of
this system by liquid alkanes.  The question of interest
concerned intercalation -- was there any tendency for
particular sizes or shapes of alkane molecules to interact
particularly favorably with these disordered monolayers --
to "complete" the monolayer or to fill it to a dense,
crystalline state?  This type of shape-selective
interaction would, of course, constitute a form of
molecular recognition.

In brief, for this system (Figure 7), we found no
evidence for this type of shape-selective wetting.  This
conclusion is not surprising, since all indications are
that shape selectivity requires the presence of molecular
cavities having defined size and rigid shape.  There is no
reason to expect such rigidity and order in a layer made
up of loosely disordered, flexible, alkyl chains.

The experiment does, however, point to one future
direction for self-assembled monolayers:  that is, toward
the development of techniques capable of assembling rigid
components into monolayers.  The ability to accomplish

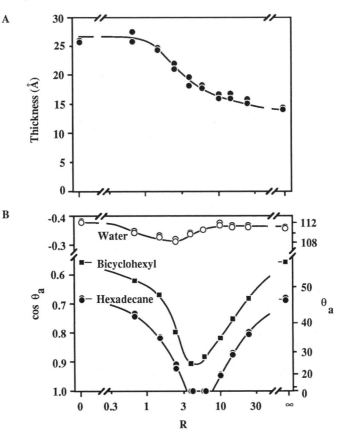

Figure 7.   Mixed monolayers on gold formed from solutions
containing mixtures of HS(CH$_2$)$_{21}$CH$_3$ and
HS(CH$_2$)$_{11}$CH$_3$.  The abscissa is the ratio, R, of
the concentrations of HS(CH$_2$)$_{21}$CH$_3$ to
HS(CH$_2$)$_{11}$CH$_3$ in solutions of ethanol.  The
upper graph plots the ellipsometric thickness
of the monolayers.  The lower graph plots the
advancing contact angles of water, hexadecane,
and bicyclohexyl.  The line through the data is
included as a guide for the eye.

this type of assembly would provide the basis for building surfaces having shape-selective cavities.

## The Future:   Does Wetting Really Have Anything to Offer Molecular Recognition?

Understanding wetting will certainly help in understanding solvation of proteins and ligands by water, and perhaps in understanding protein-ligand binding. Whether the field of molecular recognition persists for some time in its present mode -- using qualitative physical-organic models for important phenomena -- or whether it moves to a new mode based on (more) successful quantitative methods for computation and simulation remains to be seen.  If the former, results from studies of wetting will become a part of the complex process of analog reasoning, pattern recognition, and intuition used now by medicinal and biological chemists in the design of new binding agents.  If the latter, wetting will provide data from well-defined surfaces with which to test theories of solvation and potential functions for simulations.[12]  Both connections between surface chemistry and wetting and recognition are conceptually indirect. Are more direct connections possible?  I believe the answer to this question to be "yes," and propose the following topics as areas in which the fields will probably grow closer together.

1) **Structurally Complex Surfaces**.  A protein can be considered, to one approximation, as a large scaffold used to construct a cavity of molecular dimensions.[13]  This view of a protein as a rigid entity is probably largely correct in some cases (i.e., immunoglobins) and incorrect in others.  In any event, the surface supporting a self-assembled monolayer would, in principle, also serve admirably the function of scaffold.  At present, we do not know how to assemble the cavity, but we understand the principles:  Use multiple rigid components in making up the monolayer that will crystallize on the support into a thermodynamically stable, mixed, two-dimensional crystal with surface roughness appropriate to bind molecules.

At a more primitive level, one of the most versatile approaches to recognition employs partition chromatography to detect differences in recognition.[14]  The principles of self-assembly are exceptionally well suited to the rational design of sophisticated solid phases to test theories of adsorption, and, hence, of molecular recognition.

2) **Ultra-Thin Liquid Films**. It is clear for many liquids that proximity of the liquid to a solid support modifies the structure and properties of the liquid.[15] For water, one, but not the only, manifestation of this interaction is the hydrophobic effect. Self-assembled monolayers provide the opportunity for studying the interaction between organic surfaces bearing complex organic functionality and water. These studies may be carried out either using wetting or through vapor adsorption and temperature-programmed desorption.

3) **Studies Using the Force Balance**. The force balance, an instrument developed by Tabor and Israelachvili,[16] provides remarkably detailed information about the interactions between surfaces at distances of nanometers. All that is required to meld the types of studies described here using self-assembled monolayers with more quantitative studies using the force balance or, perhaps, ultimately, the atomic force microscope,[17] is the development of routine procedures for preparing suitably flat substrates.

4) **Models of Cell Surfaces**. Much of practical importance involving molecular recognition takes place at the surface of cells.[18] It is unclear how important the proximity of the receptor to the surface might be, or how surface attachment might be manifest. It is probable that many interesting aspects of receptor-ligand interactions *in vivo* will depend on the restriction of the receptor to a fluid, quasi-two-dimensional plane. Two contrasting examples of effects expected to be important are steric inhibition to binding of large ligands to receptors, due to proximity to the cell surface (and to the other components on it) and enhancement of binding of low-affinity systems (e.g. viral cell-surface receptors) due to a form of multi-point, cooperative attachment attributable to clustering of ligands by diffusion on the cell and clustering of ligands by structure on the virus.

Overall, we are convinced that much of molecular recognition is due to interaction between molecular surfaces (or more properly, interfaces), and that the study of model organic surfaces and interfaces will ultimately provide information of great value in understanding and controlling molecular recognition.

Acknowledgements

The work described here has been carried out by an exceptionally able group of coworkers, among whom are

Colin Bain, Randy Holmes-Farley, Paul Laibinis, and Kevin
Prime. Certain parts of the work are based on a
collaboration with Ralph Nuzzo (AT&T Bell Laboratories).
Support was provided by the ONR and DARPA through the
University Research Initiative, and by the National
Science Foundation through support of the Harvard
Materials Research Laboratory.

References

1)  C.B. Post and M. Karplus, J. Am. Chem. Soc., 1986,
    108, 1317. A.T. Bruenger, G.M. Clore, A.M.
    Gronenborn, and M. Karplus, P. Natl. Acad. Sci. USA.,
    1986, 83, 3801. C.F. Wong and J. A. McCammon, J. Am.
    Chem. Soc., 1986, 108, 3830. T.P. Lybrand, J.A.
    McCammon, and G. Wipff, Proc. Natl. Acad. Sci. USA,
    1986, 83, 833. J.A. McCammon, Science, 1987, 238,
    486. J.D. Madura and W.L. Jorgensen, J. Am. Chem.
    Soc., 1986, 108, 2517. J.P. Baremane, G. Cardini,
    and M.L. Klein, Phys. Rev. Lett., 1988, 60, 2152.
2)  'Contact Angle, Wettability and Adhesion', F.M.
    Kowkes, Ed., Advan. Chem. Ser. 43, American Chemical
    Society, Washington D.C., 1964. A.W. Adamson,
    'Physical Chemistry of Surfaces', Wiley-Interscience,
    New York, 1976, 3rd ed. F. Brochard, Langmuir, 1986,
    5, 432. F. Heslot, N. Fraysse, and A.M. Cazabat,
    Nature 1989, 338, 640.
3)  C.D. Bain and G.M. Whitesides, Angew. Chem. Intern.
    Ed. Eng., in press.
4)  G.M. Whitesides and G.S. Ferguson, Chemtracts, 1988,
    1, 171.
5)  Solution: C.D. Bain, E.B. Troughton, Y.-T. Tao, J.
    Evall, G.M. Whitesides, and R.G. Nuzzo, J. Am. Chem.
    Soc., 1989, 111, 321. Vapor: G.F. Ferguson, H.A.
    Biebuyck, M. Chaudhury, M. Baker, and G.M.
    Whitesides, unpublished.
6)  M.D. Porter, T.B. Bright, D.L. Allara, and
    C.E.D.Chidsey, J. Am. Chem. Soc., 1987, 109, 3559.
    L. Strong and G.M. Whitesides, Langmuir, 1988, 4,
    546.
7)  C.D. Bain and G.M. Whitesides, J. Am. Chem. Soc.,
    1988, 110, 5897.
8)  M. Wilson and G.M. Whitesides, J. Am. Chem. Soc.,
    1988, 110, 8718. S.R. Holmes-Farley, C. Bain and G.
    M. Whitesides Langmuir, 1988, 4, 921. S.R. Holmes-
    Farley, R.H. Reamey, R. Nuzzo, T.J. McCarthy and G.M.
    Whitesides, Langmuir, 1987, 3, 799.
9)  C.D. Bain and G.M. Whitesides, J. Am. Chem. Soc., in
    press. C.D. Bain, J. Evall, and G.M. Whitesides,
    ibid., in press.

10) C.D. Bain and G.M. Whitesides, Science, 1988, 240, 62.

11) C.D. Bain and G.M. Whitesides, J. Am. Chem. Soc., 1988, 110, 3665.

12) S. Dietrich, in 'Phase Transitions and Critical Phenomena', C. Domb and J.L. Lebowitz, Eds., Academic Press, New York, 1988, Vol. 12.

13) T.E. Creighton, 'Proteins: Structures and Molecular Properties', W.H. Freeman, New York, 1984.

14) S. Allenmark, 'Chromatographic Enantioseparation: Methods and Applications', Wiley, New York, 1988.

15) B.V. Derjaguin and N.V. Churaev, in 'Fluid Interfacial Phenomena', C.A. Croxton, Ed., Wiley, New York, 1985, Chapter 15; C. Tanford, 'The Hydrophobic Effect', Wiley, New York, 1973.

16) J.N. Israelachvili and A. Tabor, Proc. Royal Soc., 1972b, A331, 19. J.N. Israelachvili, J. Colloid Interface Sci., 1986, 100, 263. J.N. Israelachvili and H.K. Christenson, Physica A, 1986, 140A, 278.

17) O. Marti, B. Drake, and P.K. Hansma, Appl. Phys. Lett., 1987, 51, 484.

18) A.D. Lander, Trends Neurosciences, 1989, 12, 189. G.M. Edelman, Ann. Rev. Cell Biol., 1986, 2, 81.